Encounters in Microbiology

Volume 1

Second Edition

Collected from *Discover* Magazine's "Vital Signs"

Edited by
Jeffrey C. Pommerville, PhD
Glendale Community College

I. Edward Alcamo

JONES AND BARTLETT PUBLISHERS
Sudbury, Massachusetts
BOSTON   TORONTO   LONDON   SINGAPORE

World Headquarters
Jones and Bartlett
Publishers
40 Tall Pine Drive
Sudbury, MA 01776
978-443-5000
info@jbpub.com
www.jbpub.com

Jones and Bartlett
Publishers Canada
6339 Ormindale Way
Mississauga, Ontario
L5V 1J2
Canada

Jones and Bartlett
Publishers International
Barb House, Barb Mews
London W6 7PA
United Kingdom

Jones and Bartlett's books and products are available through most bookstores and online booksellers. To contact Jones and Bartlett Publishers directly, call 800-832-0034, fax 978-443-8000, or visit our website www.jbpub.com.

Substantial discounts on bulk quantities of Jones and Bartlett's publications are available to corporations, professional associations, and other qualified organizations. For details and specific discount information, contact the special sales department at Jones and Bartlett via the above contact information or send an email to specialsales@jbpub.com.

Production Credits
Chief Executive Officer: Clayton Jones
Chief Operating Officer: Don W. Jones, Jr.
President, Higher Education and Professional Publishing: Robert W. Holland, Jr.
V.P., Sales and Marketing: William J. Kane
V.P., Design and Production: Anne Spencer
V.P., Manufacturing and Inventory Control: Therese Connell
Executive Editor, Science: Cathleen Sether
Acquisitions Editor: Shoshanna Grossman
Managing Editor, Science: Dean W. DeChambeau
Associate Editor: Molly Steinbach
Editorial Assistant: Briana Gardell
Senior Production Editor: Louis C. Bruno, Jr.
Senior Marketing Manager: Andrea DeFronzo
Cover Design: Kate Ternullo
Cover Images: Organism photo courtesy of Rob Sleyant and Janice Capp/CDC;
  puzzle photo © Stillfx/ShutterStock, Inc.
Printing and Binding: Malloy Inc.
Cover Printing: John Pow Company

Library of Congress Cataloging-in-Publication Data
Encounters in microbiology : collected from Discover magazine's Vital Signs /
[edited by] Jeffrey Pommerville and I. Edward Alcamo. — 2nd ed.
   p. cm.
   Includes index.
   ISBN 978-0-7637-5798-4 (vol. 1) (alk. paper) — ISBN 978-0-7637-5799-1 (vol. 2)
(alk. paper)
   1. Medical microbiology—Case studies. I. Pommerville, Jeffrey C. II. Alcamo,
I. Edward.
   QR46.E54 2008
   616.9'041—dc22   2008003027

6048
Printed in the United States of America
12 11 10   10 9 8 7 6 5 4 3 2

Contents

iv   Contents

Preface

Everybody loves a good mystery! There is something about a "whodunit" that draws us in, and we are not satisfied until we have identified the villain or solved the mystery. We love to be in the shoes of the professional or amateur detective trying to solve the crime or homicide using the clues collected during the investigation.

In *Encounters in Microbiology* we follow "real life" clinicians (general practitioners, specialists, or infectious disease physicians) who find themselves in the role of medical detective (à la TV's Gregory House) investigating a "whatdunit"; that is, the villain is a mysterious infection or other associated medical malady. Infectious diseases are still with us; in fact, they account for about 25 percent of disability-adjusted life years and some 25 percent of deaths worldwide. Even in developed nations, such as the United States, each year about 3 percent of outpatients are diagnosed with an infectious disease and almost 10 percent of all prescriptions are for antimicrobial drugs.

In this revised edition of *Encounters in Microbiology, Volume 1*, the true stories originally selected by the late Ed Alcamo still represent examples of relevant medical mysteries and patient diagnoses carried out by clinicians. To make the cases more useful in the classroom, I have added a few new features and revised some others. When an ill patient presents to the clinician's office or clinic, the caregiver uses a set of steps to diagnose and treat the patient. These are described next in The Steps Used When Diagnosing and Treating a Patient. Read through them before beginning the stories, as this initial reading will provide you with the medical detective techniques needed to make the stories easier to understand and follow.

Also new to this edition are a Glossary and an Index. There are many specific medical terms (jargon) used by clinicians in their diagnosis of a patient. These terms usually identify a condition or malady affecting the patient; therefore, to help you decipher and understand these terms, a glossary has been added at the back of the book. If a term is not spelled out in the story—or even if it is—it almost certainly can be found in the glossary.

To make the stories more useful for teaching, I have revised some of the questions in the Questions to Consider section at the end of each story. Many of the questions stress critical reasoning skills. As such, they complement the epidemiological Textbook Cases found in my textbook *Alcamo's Fundamentals of Microbiology* (Jones and Bartlett Publishers). The answers to the questions are available through your Jones and Bartlett sales representative.

I wish to thank Cathleen Sether, Executive Editor, Science, at Jones and Bartlett Publishers for getting the ball rolling on the revision of this volume and Dean DeChambeau, Managing Editor, Science, also at Jones and Bartlett, for his continuing guidance and management of the project. I also want to thank Emily Haigh for her assistance in the initial preparation of the manuscript for this second edition.

As described in Ed Alcamo's Introduction, I hope you "enjoy your encounter in the world of microorganisms," and find the new additions to the medical mysteries useful and enjoyable.

Enjoy the medical sleuthing!

Jeffrey Pommerville, Ph.D.

Introduction

Microbiology can be a demanding science, with a host of new terms, a bevy of new theories, and a plethora of new insights. But it warms the soul and excites the spirit when we see practical applications for all we've learned. Placing microbiology into a useful and contemporary perspective seems to make the learning worthwhile.

And that's what *Encounters in Microbiology* is all about. In this volume, we present a series of stories involving real people and their encounters with microorganisms. In each case, the individual has reported to a hospital because he or she is suffering from a serious illness. Now it is up to the hero of the story, the physician, to find out what is going on in the patient and what is available to help. For many patients, the encounter with microorganisms ends happily, but sometimes, unfortunately, the ending is sad.

We now invite you to enter the real world of microbiology and experience real life, real people, and real cases. This volume was conceived by Dean DeChambeau, Managing Editor at Jones and Bartlett Publishers. All the articles in this volume were originally published in *Discover* Magazine, and all are true, although some of the names have been changed to protect the privacy of the patients. The articles have been written by some very talented physician-writers, and we thank them for their permission to use their stories. We also thank the editors of *Discover* Magazine for permitting us to reprint the articles.

And we thank you for giving us the opportunity to tell these stories. We hope you will enjoy your encounter in the world of microorganisms and will come to appreciate the people who fight them.

Best wishes,

E. Alcamo

The Steps Used When Diagnosing and Treating a Patient

The identification of the nature and cause of a patient's illness or disorder by a clinician is called a **diagnosis**. When the ill individual comes to the clinician's office or medical clinic, a series of diagnostic steps (**Figure A**) are set in motion. This includes a patient interview and an evaluation of the patient's reported symptoms, the physical examination findings determined by the clinician, the results of various laboratory and medical tests, and any other procedures pertinent to the investigation. If the patient's illness appears to be caused by an infectious agent, then the investigation may need to identify the causative agent and characterize the severity of the infection. When all this information has been evaluated and the clinician has reached a diagnosis, she or he can offer a **prognosis**, a prediction of the likely outcome of the disease. From this, a treatment procedure can be started and preventative (and possible public health) action initiated.

Because we are specifically interested in infectious diseases and disorders, the first step is to determine the **exposure history** of the patient.

1. Exposure History

The clinician will conduct an exposure history interview as part of the patient's overall personal and family history. Presentation to a clinician includes the patient's current illness and an oral report of his or her subjective symptoms. With regard to exposure, the clinician must cover the following topics when interviewing and examining the patient: can the patient determine the time of onset; can the patient pinpoint the time and place of exposure (e.g., home, work, recent domestic/international travel); has the patient had past infectious diseases, vaccinations, or immunological impairments; can the patient identify possible exposure sources (other humans, animals, foods, or environment).

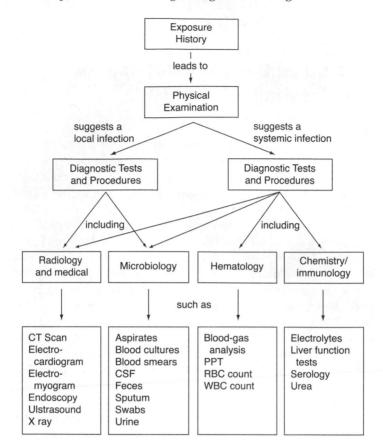

Figure A. Flow diagram of the diagnosis of an infection. CT = computed tomography; CSF = cerebrospinal fluid; WBC = white blood cell; RBC = red blood cell; PTT = partial thromboplastin time. (Modified from Greenwood, D., Slackl, R. C. B., and Peutherer, J. F. *Medical Microbiology*, 16th edition. London: Churchill-Livingston, 2002.)

The clinician's interview may indicate obvious signs or symptoms. For example, a child with a case of chickenpox (small, teardrop-shaped, fluid-filled vesicles on the torso) would be quite obvious to an examining pediatrician. In this case, little, if any, further investigation would be needed for a correct diagnosis, and the prevention of spread can be discussed. However, if this represents the beginning of an outbreak or epidemic, more specific public health measures may need to be initiated and a report made to state and national medical authorities. In other cases, signs and symptoms may not be so direct. The presence of a headache, fever, and malaise for two days could be symptoms for chickenpox as well as for a large variety of viral and bacterial infections.

In the interview with the patient, a review of body systems may uncover other parts of the body being affected by the disease. For example, coughing and shortness of breath may indicate respiratory system involvement. On the other hand, a burning on urination would suggest a urinary tract infection. All are part of the disease detective work done to locate the physical site of symptoms or what is called an **anatomical diagnosis**. Such a diagnosis may allow the clinician to narrow the list of possible infections or infectious agents.

2. The Physical Examination

As part of the **physical examination**, the clinician does a systematic examination of the patient, taking a blood pressure reading, measuring heart beat, and measuring body temperature. Specific emphasis is placed on the part of the body affected by the illness. For example, the throat, chest, and lungs are examined if a respiratory system infection is suspected.

The examination may allow the clinician to narrow the possibilities of diseases and/or infectious agents that would fit the clinical findings. The exposure history and physical exam may lead to the determination if the infection is local, such as the lungs or urinary tract, or systemic, involving several tissues/organs in the body. The clinician may then be ready to make a **differential diagnosis**, which narrows down the potential diseases to just those few that fit the clinical findings. If the presenting symptoms in a 45-year-old patient are a three-week cough, fever, chest pain, and coughing up blood or sputum, a differential diagnosis may include several respiratory infections but primarily tuberculosis (TB) as the cause. On the other hand, if the patient remembers a tick bite and has an expanding red rash at the bite site, the differential diagnosis is almost certainly Lyme disease, as few other arthropod-borne diseases have these specific signs.

3. Diagnostic Tests and Procedures

As illustrated by many of the stories in this book, the first two stages—exposure history and physical examination—are carried out rather quickly, often on the initial interview.

However, the clinician may order one or more specific diagnostic tests to narrow down the short list of possible infections. Such tests or procedures can take some time and may be expensive. This is one reason why a clinician might attempt a final diagnosis through the physical examination or by using a minimal number of "standard" tests. For example, if TB is suspected, a chest x-ray or tuberculin skin test may be ordered to confirm the diagnosis. In addition, some diagnostic procedures may be noninvasive while others are invasive. Thus, comfort to the patient must be considered when diagnostic tests or procedures are being considered. Also, if someone such as a general practitioner is treating a patient, she or he may consult with an infectious disease specialist to obtain an expert opinion. Even Google searches have been used to help with patient diagnoses!

In some cases it may not be necessary to identify the specific pathogen as part of the diagnosis. For example, if all the diseases or agents identified by a differential diagnosis would be treated in the same way, or not at all (e.g., common cold, measles), there probably is no need to identify the pathogen. On the other hand, sometimes it is necessary to identify the actual causative agent of the infection. This **etiological diagnosis** may be important especially if it is a particularly dangerous disease. Again, taking TB as an example, drug-resistant TB is increasing rapidly worldwide. Therefore, it might be necessary to determine the drug resistance of the particular TB strain infecting the patient. In this case, a sputum sample would be taken for culture and growth of the bacteria. Then an antibiotic resistance determination would be made of the bacterial strain.

During the differential and etiological diagnoses, the clinician needs to be aware of a number of important epidemiological issues, especially if it is a disease like TB. The clinician needs to know if other individuals are at risk. Do particular behaviors (traveling, working in crowded places, etc.) expose one to the disease, and has the patient engaged in these behaviors? What is the geographical distribution of the disease, and has the patient been in these locales? Have there recently been additional cases reported locally? Has the patient been immunized, if possible, against this disease? During the patient interview and examination, many of these questions may be answered by the patient, assuming (which one often cannot) that the patient's ability to "self-report" is honest and accurate.

Although diagnoses and diagnostic tests obviously demand good judgment on the part of the clinician, for some conditions, written flow diagrams called **decision trees** (or algorithms) exist for making diagnostic decisions and for treating the patient; in other words, "If the patient has this, do the following test." Medical and health insurance companies often use diagnosis and treatment algorithms. An example of a decision tree for evaluating a suspected TB patient is illustrated in **Figure B**. As you can see, they can be quite extensive.

4. Treatment

Once the clinician has reviewed all the clinical information and diagnostic tests, hopefully a correct diagnosis can be made and the prognosis issued. Then treatment can begin. Note: Often as a precautionary measure, treatment may begin while diagnostic tests are being run.

There are two possible treatment scenarios. In **symptomatic treatment**, the clinician treats symptoms, such as pain, fever, cough, or muscle aches accompanying the underlying disease. Pain relievers, antihistamines, or cough suppressants may be prescribed for something like a cold or the flu. These treatments are simply supportive, making the patient feel better without influencing the final outcome or progression of the disease.

In a **specific treatment**, the clinician is specifically treating the diagnosed disease and hopefully affecting the final outcome. Typical specific treatments might be prescribing an antibiotic for a sinus infection or several antibiotics for something like TB. Antiviral agents, such as acyclovir, might be prescribed for shingles, although in actuality it is only treating the symptoms. Hopefully through proper treatment, the patient will progress through a period of disease decline and complete convalescence. However, less optimistic outcomes due to deadly pathogens sometimes occur, as some of the stories will describe.

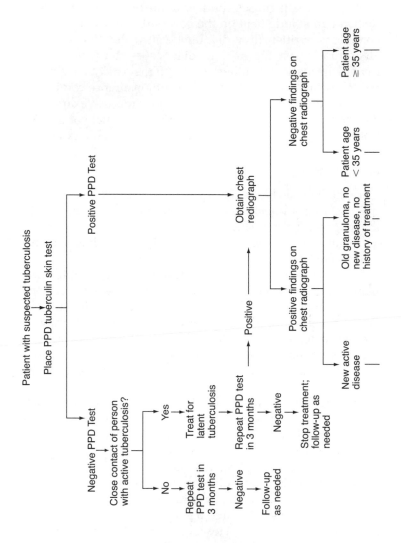

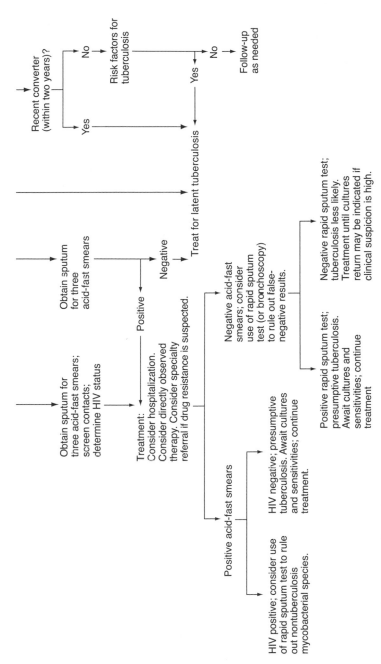

Figure B. Flow diagram for evaluation suspected tuberculosis patients. PPD = purified protein derivative; HIV = human immunodeficiency virus. (Modified from Jerant, A. F., Bannon, M., and Rittenhouse, S. Identification and management of tuberculosis. *American Family Physician* 61(9): 2667–2682, 2000.)

Depending on the signs and symptoms, a differential diagnosis of a patient may initially include meningitis, HIV infection, Lyme disease, food poisoning, or a host of other illnesses. When the final diagnosis is returned, however, it could be unlike anything anticipated. Moreover, as this encounter illustrates, the source of infection may never be definitely identified.

FEVER WITHOUT A CAUSE

PAUL ARONOWITZ

"My boyfriend needs some more codeine for his headache," the caller said. Even over the phone, the edge in her voice told me something was terribly wrong.

"Would you start at the beginning?" I asked.

The woman took a deep breath and began. "Two days ago my boyfriend came to your emergency room. He'd been having terrible headaches and fevers for five days before that. His temperature was 104 when the doctor saw him, so she thought he might have an infection in his head .... "

"Meningitis?"

"Yes. She did a spinal tap so she could get some fluid out of his spinal column to see if it was infected." So far all seemed in order. A man with a fever. A man with a headache. Meningitis might explain both symptoms, and it might kill him if it went untreated.

"Then she told us that Timothy didn't have meningitis. She said he had a bad virus and that his fevers and headaches would go away over the next few days."

"And?"

"They haven't. He's getting worse. He's sitting here shaking, his teeth chattering. He's drenched in sweat and has a terrible headache. It's been over a week now. Could you please call in a prescription for some codeine?" The edge returned to her voice. But I was a bit skeptical of the scenario Timothy's girlfriend had described; it would be somewhat unusual for a viral infection to make someone sick for so long. And not only was Timothy not getting better—it seemed he was getting worse.

"Put him on the phone," I told the young woman. Timothy greeted me a few moments later, sounding tired.

"Forget about the codeine," I said to him. "We need to find out why you're still having these fevers." He grunted in agreement. "Come back to the emergency room right now."

Within an hour Timothy and his girlfriend arrived. He was 36 years old, ruggedly built, with chiseled features and chestnut brown hair. His pale face was partially covered with several days' worth of beard, and his shirt was drenched in sweat. His temperature was 101.

"I think my temperature's broken for the time being," he said quietly. "It comes twice each day—early in the morning and again in the evening. I take some aspirin and sweat it out. My teeth rattle like mad, and sometimes the whole bed seems to be shaking with me." He was describing what doctors call rigors—the severe, shaking chills that accompany high fevers.

I asked him to tell me his story again, then repeated it back to him to make sure I'd gotten all the details right. This was one of the things I loved about medicine: the search for details that, while they might seem completely mundane to a patient, when applied correctly might solve the riddle of his or her illness.

Timothy sat back in his chair, leaning his head against the wall. He told me that until a week earlier he had been in perfect health. He remembered the first moment he felt sick: he was walking to his car with his girlfriend after a meal at an Italian restaurant. A chill came over him and his teeth began to chatter. He went home, his head pounding, and stayed in bed for the next four days. His headaches came and went with his fevers. Then, two days ago, he had come to our emergency room.

At first I suspected endocarditis, an infection of the heart valves that can be caused by infection during dental work or by injecting drugs with dirty needles. One of its symptoms can be intermittent fevers. I asked whether Timothy had ever used intravenous drugs or had had any recent dental work. No, he answered gruffly. And he denied any risk factors for HIV infection. Thinking he might have contracted some animal-borne infection, I asked if he had any pets. But Timothy said he didn't have any pets, and he assured me that he hadn't been hiking in woods or fields where he might have been bitten by ticks. So tick-borne diseases, such as Lyme disease or Rocky Mountain spotted fever, seemed out of the question.

When I asked if he had traveled abroad recently, he said no. He mentioned that he had gone to Mexico six months earlier and to Thailand a year earlier, but he hadn't become sick during either trip. It seemed unlikely that a parasitic infection or bacterial disease contracted abroad would manifest itself so dramatically 6 or 12 months later.

I then reviewed his dinner the night he got sick. Spaghetti and meatballs, salad, gelato; two glasses of red wine; a cup of coffee—decaffeinated. His companions had eaten the same dishes and drunk the same wine; all were healthy. So food poisoning was unlikely.

When I asked about possible exposure to illnesses on the job, he told me he was a freelance writer. Little risk of febrile illness there.

I examined him thoroughly, listening long and hard to his lungs, heart, and abdomen. I pushed everywhere, pulled on his extremities, twisted his fingers, looking for some sign of the swelling and tenderness that can signify joint disease or an autoimmune disease such as lupus. I peered into his nose, mouth, and ears. All seemed in order.

Despite our growing reliance on expensive medical technology, common illnesses can usually be diagnosed simply by listening to a patient's story, obtaining a medical history, and performing a physical exam. But in Timothy's case, technology was all I had left. I began with the basic tests, sending Timothy to radiology for a chest X-ray and then to the laboratory for routine blood and urine tests for bacterial infection. His chest X-ray was normal—no sign of the haziness that fluid or inflammation in the lungs can produce. His urine was clean—free of bacteria and white blood cells—and his white blood count, which infection commonly elevates, was also normal.

I continued seeing patients while Timothy waited patiently for the lab test results. His fever had subsided and he had begun to look and feel better. Though a few more lab tests were still being performed on his blood, I decided to let him go home, where he would be more comfortable. But I made him promise to return the next morning, and I assured him that we would discover the cause of his fevers. While an unusual virus might still be at the root of his troubles, I was convinced that something else was going on. But I didn't know what.

Timothy thanked me for my efforts and headed out the door. I sat down at the computer and began reviewing his lab results. They were normal—I hadn't missed anything. Even technology was failing me.

A moment later, the last of his blood tests appeared on the computer screen. These tests screened for enzyme activity in the liver, and they were slightly elevated. The elevation was so minimal, in fact, that I wouldn't have paid attention to it in another patient. I stared at the screen for several minutes. That slight rise might indicate inflammation in the liver. These slightly abnormal blood tests were the only clue I had to the cause of Timothy's fevers.

I leaped out of my chair.

"I'll be right back," I said to one of the other ER physicians. I rushed out of the ambulance bay doors and jogged toward the parking lot. At the far end of the huge lot I saw a head of chestnut brown hair slowly making its way down an aisle. I caught up with Timothy just as he and his girlfriend were getting in her car.

"You need to come back," I said. His girlfriend rolled her eyes. Twice we had failed to diagnose Timothy's illness, and here I was chasing them through a parking lot just after telling them to go home.

I tried to explain about the liver tests. They looked skeptical. "It's all we have," I pleaded. "I want you to stick around a little longer so that we can have the radiologists perform an abdominal ultrasound. They'll use sound waves to get a look at your liver and gallbladder, to see if there's something there."

"Like what?" Timothy asked, tilting his head to one side.

"I'm not sure. I'm groping, but it's all we have to go on." I looked into the eyes of a young man who had barely been sick a day in his life. Like most people suddenly struck ill, he just wanted an answer, a cure, and to get on with his life.

"Look, I'd really like to get home and get some rest," he said, eyeing his girlfriend's car. "About 14 hours of sleep would feel good right now. I'll come back in the morning."

"It's important you come back now," I repeated. I felt foolish for having let him go before all his lab tests were back. Now I stood in the parking lot begging him to return for more tests that still might not provide an answer. He stared at me, then glanced over at his girlfriend. He shrugged, pivoted, and began walking slowly back toward the emergency room.

One hour after I'd sent Timothy up for his abdominal ultrasound, the radiologist called. "I hope you put this guy's admission papers in," he said, "because he's a keeper if ever I've seen one."

I swallowed hard.

"What did you find?"

"There's an abscess the size of a volleyball in the right lobe of his liver, one of the biggest I've ever seen," he replied calmly.

Timothy returned to the emergency room a few minutes later. I put him on a gurney, and a nurse started an IV line.

"What's that thing in my liver from, Doctor?" Timothy asked.

"Apparently it's filled with fluid, probably pus. It could be a bacterial infection that spread from your colon into your liver or spread from somewhere else in your body." I thought about his travel—a trip to Mexico six months earlier and Thailand a year earlier. "Or it could be from an infection that you picked up in another country."

That evening, Timothy spiked another high fever and looked extremely ill. Because the test results identifying the infecting organism would not be back for several days, the doctors caring for Timothy assumed he had a bacterial liver abscess and ordered that a catheter be put into his liver to allow it to drain. They also started him on an antimicrobial drug that would clear any parasitic organisms. With Timothy as ill as he was, compulsivity might payoff.

Two days later I stopped in to see Timothy. He looked much better and tossed a friendly smile my way. He said his catheter was going to be pulled out that day.

"You were persistent," he observed, "and I thank you for that."

"I only wish we had a better idea of what caused this thing," I told him.

"Amoebas is what they tell me," he said, chuckling. "They think I might have picked them up in Mexico six months ago. Until now, I didn't have any idea I was infected."

When Timothy's blood tests finally came back, they were positive for exposure to Entamoeba histolytica, an amoebic parasite that feeds on red blood cells in the human gut. The parasite passes from host to host in capsules shed in the feces. People tend to become infected when they ingest these capsules, either through fecal contamination of water or food or by direct fecal-oral contact. After the capsule is eroded in the intestine, the parasite invades the intestinal lining, usually causing diarrhea, abdominal pain, and bloating. This form of infection, called amebiasis, is common in Mexico and other developing countries. Something as simple as a garden salad might have been the source of Timothy's infection.

But people suffering amoebic infections can sometimes be symptomless for months. And in rare cases, the amoebas can spread from the colon to the liver, as they had in Timothy. When this happens, the body mounts an immune response to try to wall off the organism and prevent it from spreading to other tissues. The result is a pus-filled abscess.

In most cases of amoebic abscess, patients develop symptoms within three or four months. But Timothy was his own patient, not a statistic, and it had taken more than six months for the abscess to make him ill. Even more surprising was that the huge abscess hadn't provoked any tenderness over his liver. Presumably it had grown so slowly that it hadn't caused any noticeable pain.

As is so often the case, the answer to a puzzling ailment had been in the patient's medical history. I realized now how vital Timothy's travels outside the country were to understanding his mysterious fevers. Fortunately, Timothy was recovering rapidly and would be discharged in a few days. When I stopped by for a last visit, he bore no resemblance to the weakened, haggard man I had met in my ER room. I looked at him, restored to health by a simple antimicrobial treatment, and smiled.

UPDATE

The genus Entamoeba *consists of single-celled protozoal parasites that can infect vertebrates. The microorganisms have a simple life cycle, consisting of an infective cyst stage and a multiplying tropho-zoite stage. As probably occurred in this encounter, transmission occurs through the ingestion of cysts in fecal-contaminated food or water, which release trophozoites in the host's digestive tract. Globally,* Entamoeba histolytica *ranks third among the leading causes of morbidity and mortality due to parasitic disease in humans (after malaria and schistosomiasis). An estimated 50,000 to 100,000 individuals worldwide die every year from amoebiasis.*

QUESTIONS TO CONSIDER

1. What diseases were considered as part of Timothy's recent medical history?

2. What precautions might Timothy have taken in Mexico to reduce the chances of an infection?

3. Why did the liver tests point to the possibility of disease, and how was the final diagnosis eventually made for this patient?

4. Suppose this disease had occurred in an elderly person whose immune system is weakened. What might the result have been? Why?

5. Why wasn't the source of Timothy's disease identified?

As this encounter illustrates, it may take only a few crucial observations for a physician to return a preliminary diagnosis for an infectious disease. Then the race is on to confirm the diagnosis and institute life-saving therapy, hoping that the correct decisions have been made. The experience and insight of the physician are determining factors in the patient's return to normal.

FIRESTORM

TONY DAJER

It was 7 A.M. when I arrived at the hospital. The soft blue dawn surrendered to the emergency room's fluorescent, round-the-clock glare. Reuben, the night-shift attending physician, smiled blearily.

"How'd it go?" I asked.

"Oh, the usual. And nothing to sign out."

My turn to be thankful. But now another emergency: coffee. I was heading back to the doctors' offices when a faint mewling began outside. Ambulance backing up. I gulped a mouthful of coffee and headed out. The ambulance bay doors hissed open. A young woman's voice came howling in: *"Leave me alone! Ow ... oh. Damn you! Leave me alone!"*

Bleed, I thought, though drug overdose would have been the better bet. I was working in an emergency room serving a small northeastern city. The city's economy had crashed with the Atlantic fish stocks, and drug abuse and psychiatric problems abounded. But the worst mistake in medicine is to label patients wacky before you label them sick. And nothing makes someone delirious as shockingly fast as a hemorrhage in the brain.

Four stout paramedics pinned the girl down. Ann, the head nurse, and I converged on the stretcher. The nearest medic filled us in.

"Virginia Carlson, 22-year-old, horrendous headache at 4 A.M. Vomited twice. Then, the parents say, she got like this. Been fighting the flu for a couple of days."

"*No!* Go *away*," the girl hollered. "*Ow … ow. I hate you!*"

The medics hung on so we could tie soft restraints around her wrists and ankles. A brain hemorrhage was still possible, but the flulike symptoms described by her parents suggested a new contender: meningitis, inflammation of the membranes enveloping the brain and spinal cord that is usually caused by a bacterial infection. While Ann, two more nurses, the four paramedics, and Bob, our 250-pound nurse's aide, wrestled Virginia into restraints, I found her parents.

Mrs. Carlson was astonishingly calm, given that her daughter was being restrained by an ER crew. Mr. Carlson, a thin, quiet man, took his cues from his wife.

"What do you think is wrong, Doctor?" Mrs. Carlson asked.

"I can't say yet, but I need to ask some personal questions very quickly."

"Please."

"Does your daughter use *any* kind of drugs or drink alcohol?"

"No, never," came the firm reply. "I know she's used marijuana now and then, but she's a sweet, responsible, wonderful kid."

"Medications? Even over-the-counter?"

"No."

"She hasn't been under more stress lately, she's never taken pills or tried to hurt herself?"

"No."

"And no penicillin allergy? She's taken it before?"

"No allergy and yes, she's taken it."

Just then, Ann called out: "Rectal temp 101."

"Be right back," I said. The crew at the stretcher had thinned. In her delirium, Virginia struggled against her restraints, desperate to flee the demons she was imagining had bound her hand and foot.

Ann came over.

"You thinking what I'm thinking?" she asked. "Meningitis—although there's still a small chance of a bleed, even with the fever," I answered. "But what she needs right now is penicillin—lots and fast. Four million units. Repeat in four hours."

A single dose of 1.2 million units of penicillin will cure strep throat. Virginia needed 20 times that much per day every day until she cleared the infection.

"And after the penicillin?"

"CT scan first, then spinal tap."

"Sedation?"

"Tons. She needs to lie still for both. We can't just pin her down."

In treating first and diagnosing second, I was gutting medical dogma, but meningitis demands a shoot-first-ask-questions-later approach.

The meninges make up the protective wrapping of the brain and spinal cord. There are three layers of meninges, running from forebrain to spinal cord, which is why there is no such thing as exclusively "spinal" meningitis. The innermost membrane—the meshlike pia mater—hugs the surface of the brain and spinal cord like shrink-wrap. The outermost membrane is the dense dura mater. Sandwiched between is the arachnoid, a loose-fitting membrane that holds the cerebrospinal fluid, or CSF. This fluid-filled compartment insulates the fragile brain from damaging jolts, and it is here, in the warm CSF, rich in glucose and proteins, that meningitis-causing bacteria can flourish.

For an ER doctor, it is uncomfortably easy to mistake meningitis for drug-induced dementia. Thankfully, meningitis can be confirmed by performing a spinal tap, in which some CSF is drawn out through a long needle inserted into the lower back and is then examined for bacteria and infection-fighting white cells. But before performing a spinal tap, a doctor must rule out the possibility of a tumor or broken blood vessel in the brain. Both conditions can create dementia—as well as dangerously high pressure within the brain. When pressure is greatly increased inside the skull, drawing CSF out of the spinal canal can force the brain through the foramen magnum, the opening in the skull where the brain meets the spinal cord. The soft, gelatinous brain tissue is then crushed against unyielding bone. Fortunately, a CT scan, which can detect tumors or brain hemorrhages, can head off such a disaster. But CT scans take time. And meningitis won't wait.

If a patient is battling an infection as life-threatening as meningitis, antibiotics should be given immediately. Unfortunately, doctors are often reluctant to fire such heavy antibiotic artillery without having lab results that prove infection. So some doctors wait for a cell count before starting life-saving antibiotics. In addition, doctors want to make an accurate diagnosis, and if they give antibiotics before they do a spinal tap, the lab culture will turn up negative.

The way around this is to give antibiotics promptly and then use tests that detect antibodies to bacteria rather than the bacteria themselves. In a healthy young woman like Virginia, I knew there could be only two major contenders, and both bacteria, the pneumococcus and the meningococcus, are vulnerable to penicillin.

The drug dripped into Virginia's IV line. Fifty minutes had passed since Virginia arrived, and it had taken that long to get her restrained and prepared for treatment.

I prayed the delay wouldn't make a difference.

Ann brought Virginia's parents over. Mrs. Carlson began to stroke her daughter's forehead.

"Did you see the faint rash on her chest, Doctor?" she asked. "Yesterday she was taking a warm bath for the muscle aches and said, 'Mom, what do you think these red spots are?' "

At that point, *all* my alarm bells went off. In all the uproar, we hadn't yet stripped Virginia down and done a skin exam. Ann pulled back the sheet. Virginia's chest and abdomen were speckled with little red bumps. Even more deadly than meningitis is meningococcemia, a condition in which meningococcal bacteria spread through the bloodstream. Meningococcemia is one of the fastest infectious killers around, and it's known for its peculiar rash. Outpacing even the latest made-for-TV horror virus, it can dispatch its victims within a few hours of symptom onset. I suddenly wanted those 50 minutes back very, very badly.

The meningococcus bacterium, technically known as *Neisseria meningitidis,* is not some rare imported microbe that catches an immune system off guard. Rather these bacteria are so common that many of us harbor them in our noses and other mucous membranes. Fortunately, antibodies that stand sentry like saguaro cacti on our mucous membranes usually snag any invaders.

But in rare cases a virulent strain of *N. meningitidis* manages to overwhelm the sentinel antibodies and penetrate mucosal cells in our respiratory tract. In a feat of Machiavellian biochemistry, they somehow masquerade as a familiar cellular import, duping the cells into swallowing them up in membrane-bound capsules and ferrying them to capillaries beneath the mucosal barrier. Once in the bloodstream, they deceive the immune system by displaying the same molecular ornamentation as red blood cells. Thus disguised, they slip through the blood-brain barrier, which normally protects the brain against infection, and run riot in the defenseless CSF. The result is meningitis. Of the estimated 20,000 cases of bacterial meningitis in the United States each year, roughly 2,500 will be meningococcal infections.

In meningococcemia, *N. meningitidis* multiplies in the blood, causing so much infection that the immune response can swerve out of control. In a normal immune response, the walls of the blood vessels become more porous to permit immune cells into infected tissue. But in overwhelming infections, the blood vessels leak so much that blood pressure plummets. Blood pours into the skin and the internal organs, causing massive hemorrhaging.

When there is *severe* internal bleeding, a blotchy, purplish rash sometimes covers the skin. Fortunately, Virginia's rash didn't look like that. But the Carlsons needed to prepare for the worst.

"She has meningitis," I told them. "It is also possible the infection has spread through her bloodstream."

"Could she die?" Mrs. Carlson asked, without a quaver in her voice.

"Yes. The critical period will be the next 24 hours. Her blood pressure is stable now. If it holds like this till evening, we'll be miles ahead."

"If she survives, could she have problems? Brain damage?" Mrs. Carlson looked me right in the eye. She wanted the truth.

"There's a chance," I answered hesitantly. "Yes."

Fortunately, Virginia's CT scan was negative. No tumor or hemorrhage in the brain. With Bob's bulk blanketing Virginia, and Ann hugging her into a C shape that would expose the spaces between her vertebrae, I performed the spinal tap. I slipped the needle between two vertebrae and extracted the spinal fluid. It was cloudy, as I expected, but the pressure I detected added to my fears. In a normal patient, the pressure in the spinal canal will not force the CSF above 180 millimeters on the spinal tap manometer. In Virginia's case, the pressure exceeded 550—the instrument's highest measure.

That astronomical pressure meant Virginia was fighting an enormous brain infection. When the body fights infection, immune cells gather at the infection site, causing swelling. Virginia's brain was so swollen that she needed to be managed by a neurosurgeon, which our hospital didn't have.

The lab called. Virginia was infected with meningococcus. Unlike pneumococcal bacteria, which tend to cause only isolated cases of meningitis, a virulent strain of meningococcus can cause outbreaks of meningitis in otherwise healthy young people. The bacteria are spread by close contact or through a sneeze or a cough; outbreaks tend to crop up in the close quarters of schools or military camps. For unknown reasons, some people suffer invasive infection, while others remain healthy carriers.

Who had had close contact with Virginia? I ticked off all the possibilities: her family, the medics who had wrestled her down, the ER staff, and friends who had spent the night at her house two days before. They would all need a two-day regimen of the drug rifampin, which kills meningococcus in the nose and throat and eliminates any risk of disease. Then there were the casual contacts. Because of their limited exposure, they wouldn't need—but would understandably clamor for—the antibiotic treatment.

We arranged Virginia's transfer to a nearby hospital. By the time she left she was unconscious—an effect, I hoped, of the sedatives we had given. Ann took along a bag of penicillin to give her en route.

"Virginia's blood pressure hasn't moved a millimeter," I told the Carlsons as they prepared to accompany their daughter. That was good news. When Virginia came in, her blood pressure had been normal. If it was holding steady, the infection was probably not spreading through her bloodstream.

The next morning, Ann greeted me with a rueful look.

"They just took Virginia's brother into the emergency room. Headache and chills."

Oh no, I thought. "We covered him with rifampin, didn't we?"

"You bet," Ann replied.

"Whew. Is the town going nuts?"

"What do you think?"

"And how's Virginia?"

"She's still unresponsive. The doctors at the other hospital don't sound too optimistic."

My heart sank. Fifty minutes. What had taken us so long?

Still, it was too early to tell.

That afternoon came the verdict on Virginia's brother: his spinal tap was negative. It was just the flu.

Two days later the clouds parted a bit.

"Virginia opened her eyes this morning," Ann told me. She would live. But what about residual brain damage? It can take months for brain injury to show up.

Four weeks later, I called Virginia's mother.

"Oh, Virginia's back at work now and doing pretty well, Dr. Dajer. The worst thing is the headaches—they make her think the meningitis is coming back. They scare her."

I knew Virginia was probably out of the woods. She wasn't showing any signs of brain damage, and reinfection with the bacteria was pretty unlikely. But I also knew what Virginia was feeling. Meningitis is a terrible foe, and no matter how fast a doctor acts, it can act faster. Virginia had been lucky.

UPDATE

Neisseria meningitidis *remains a leading cause of bacterial meningitis and sepsis in infants and teenagers in the United States. Although the incidence is declining [there was an all time low of 219 cases reported to the Centers for Disease Control and Prevention (CDC) in 2006], survivors still face serious health sequelae, including hearing loss, amputations, and, as occurred in this encounter, severe headaches or cognitive impairment. A new meningococcal vaccine called Menactra® was approved by the United States Food and Drug Administration (FDA) in January 2005 for use in individuals 11 to 55 years of age. The CDC believes the new vaccine, which is longer lasting than the older meningococcal vaccine, will substantially reduce the burden of meningococcal disease in the United States.*

QUESTIONS TO CONSIDER

1. In the few short moments after Virginia's arrival at the hospital, the physician in this encounter made some key observations leading to the correct diagnosis. What were they?

2. What did the presence of Virginia's flu-like symptoms during the previous days indicate?

3. Concerning Virginia's brother, the physician recalled, "We covered him with rifampin, didn't we?" What did he mean by this statement?

4. Why is "treating first and diagnosing second" important in this instance?

5. Why is meningococcemia more dangerous than meningitis? How do the two conditions differ?

A rigid neck, a rigid torso, and tight facial muscles can accompany numerous conditions, including psychosis and drug overdose. As this encounter describes, they can also be signs of a serious bacterial disease.

BLINDSIDED BY TETANUS

CLAIRE PANOSIAN DUNAVAN

Eduardo rubbed his jaw and tried to open his mouth, wondering about the tight muscles in his face and neck that had plagued him all day. Then he noticed the flashing lights of a police cruiser in his rearview mirror. As an illegal alien in a battered pickup without cash, driver's license, or friends, Eduardo felt that this was becoming his worst nightmare.

Charged with weaving across lanes and driving an unregistered vehicle, Eduardo spent the next two days in a holding cell. As the hours passed, his cell mates noticed that he grew stiff, grinned oddly, and ignored his food. Then, one of the guards saw him violently jerk his neck and torso. The guard thought, "This guy's faking seizures to get out of jail." But Eduardo's spasms persisted, and other prisoners began backing away from him. The staff decided to pack him off to the county hospital's psychiatric unit.

During my years as the sole infectious diseases specialist at that small county hospital in southern California, I wasn't called to the psychiatric emergency room often. But when I was, the cases were never boring—like the woman with obsessive-compulsive disorder who swallowed nails, tacks, and the metal springs from ballpoint pens. She came in with a fever and a boardlike abdomen—a textbook case of peritonitis due to a perforated intestine. But that's another story.

Eduardo posed a challenge. As I and the resident both knew—but the police did not—psychosis and overdose were not the only conditions that could produce a rigid neck and torso, a mute smile, and jerking movements. An infection of the central nervous system was another possibility, and we'd recently seen a few cases of mosquito-borne encephalitis in the area.

"*¿Como esta?*" I asked as I approached the young man lying on a gurney in a curtained cubicle. The greeting was a courtesy. Eduardo was in no shape to talk. Invisible pulleys had stretched his mouth into a tight smirk. But his eyes were wide open, alert, and terrified—no sign of confusion or coma.

"Great—you got here fast!" The resident's voice rang out as he flung back the curtain.

The sharp sound and sudden motion startled Eduardo. His head jerked back, his shoulders and trunk arched up, and he gasped in pain. But he remained conscious throughout the 15-second attack. That's not consistent with spasms induced by brain disorders. This was no ordinary seizure.

Suddenly the diagnosis dawned on me. Twelve years earlier, as a medical volunteer in Haiti, I had watched a rigid yet fully conscious pregnant woman arch her body in just the same way.

She'd had tetanus.

"Get the ICU team here as soon as possible," I said to the resident. I spoke softly to avoid startling Eduardo into another spasm. "The next time this happens, he could stop breathing," I told the resident. "You make sure he gets an airway. Meanwhile, I'll order up some antitoxin."

In the specialty of infectious diseases, few physical displays are as dramatic as the spasms provoked by tetanus. Its cause is a protein toxin so potent that many victims require months to recover from its effects, if they survive at all.

But the toxin is not the ultimate perpetrator of tetanus. That honor is reserved for the bacillus *Clostridium tetani*, which produces the toxin. Excreted in the feces of animals and widely distributed in soil, mature *C. tetani* resemble tennis rackets, bulging at one end with a hardy spore. It doesn't always take an old nail puncturing a foot to get these into a human host. All the bacteria need is a minor breach of the skin—a laceration, a burn, or even an insect bite. And if they land in tissue that receives little oxygen, they will thrive—multiplying and manufacturing their deadly product.

Once secreted, the toxin molecule sneaks into the rootlike hairs of nerve fibers, climbs toward the spinal cord, and binds itself to inhibitory neurons, thus disrupting their function. That takes the brakes off the peripheral nerve cells, and they start firing faster. The result is muscle rigidity that typically begins in the head and neck, then moves to the chest and abdomen, and eventually reaches the extremities.

Lockjaw, or trismus, is an early sign of tetanus. It means the toxin has affected nerves in the masseters, or chewing muscles. Another early symptom is *risus sardonicus*, a term from Roman times for the tetanus victim's telltale smile, raised eyelids, and wrinkled forehead. The most vivid hallmark of all is the wrenching spasms, which result when two opposing muscle groups are simultaneously activated. The spasms can be triggered by anything from a sudden noise, movement, or draft of air to such internal stimuli as a full bladder or a cough.

Fortunately, most people in industrialized countries needn't worry that everyday scratches and scrapes will yield an internal harvest of tetanus toxin. Because they've received a series of tetanus vaccines in childhood as well as the occasional tetanus booster, their bodies have plenty of protective antibodies. Reported tetanus cases in the United States often number no more than 100 a year.

But people in the developing world are less likely to receive tetanus vaccines and they suffer the consequences. Tetanus kills an estimated 300,000 each year; almost all deaths occur in developing countries. Newborns are particularly vulnerable. During the first few weeks of life, their only defense against pathogens comes from antibodies imported from their mothers' breast milk. Infants born to nonimmunized mothers are tetanus cases waiting to happen. One dirty knife or soiled bandage on the umbilical stump is all it takes. Today neonatal tetanus accounts for over half of the more than 500,000 cases worldwide.

In my quick exam of Eduardo, I hadn't seen a scratch. I suspected tetanus, but there's no definitive diagnostic test for the disease because the toxin hides away in the central nervous system. To confirm my suspicion, I needed to exclude the possibility that another condition was mimicking tetanus symptoms.

Tests of Eduardo's electrolytes were normal, which ruled out a low calcium level as the cause of his spastic muscles. And Eduardo's spinal fluid showed no signs of infection; that ruled out encephalitis or meningitis. And just in case he was suffering from dystonia—a movement disorder triggered by certain prescription drugs—he got a dose of diphenhydramine (Benadryl), the usual antidote. That maneuver proved fruitless as well. The only remaining tests were blood and urine assays for strychnine, and those results might not be back for days. Tetanus was the leading contender.

"We'll start the antitoxin as soon as pharmacy brings it up," said the ICU chief, taking me aside. "In the meantime, he's intubated, with diazepam [Valium] by IV. Now what about antibiotics?"

Although Eduardo had no visible signs of infection, at least somewhere in his tissues there must be *C. tetani* pumping out toxin. Penicillin was in order. The drug would wipe out the toxin-producing bacteria. And we hoped the antitoxin—antibodies culled from horses or humans immunized against tetanus—would intercept the poisons in his blood and prevent his symptoms from getting worse.

Unfortunately, its effects were far from Lazarus-like. Eduardo remained in the ICU for a full month, while the toxin was slowly leached from his spinal cord and brain. I was hoping for a full recovery, but sometimes tetanus so damages nerves that muscles are left permanently weakened. Even muscle relaxants, low lights, and tiptoeing doctors and nurses couldn't prevent Eduardo's spasms, so we paralyzed his muscles and put him on a ventilator. Thankfully, he made it through.

Several weeks after his discharge from the hospital, I saw Eduardo at a follow-up visit. He was still thin and leaning on a cane. When I greeted him in the hall, he seemed to remember me.

"Tetanus vaccine?" he responded laconically to my first eager question. "I don't remember any vaccines in the village where I grew up."

I made a mental note to ask our nurse to vaccinate him. Ironically, so little toxin is released during an infection that even a full-blown case of tetanus builds no immunity against future attacks.

"What about an injury?" I persisted. "Usually a wound precedes tetanus."

"Ah, the soccer game," he mused. "A few weekends before I started getting stiff, something sharp went right through the sole of my shoe. Glass, I think."

I had one last question. "What was it like when you knew you were sick, but everyone else thought you were crazy?"

"Sorry, he can't talk about that," his brother piped in. "Attorney's orders."

I smiled. Only in America.

UPDATE

The number of reported cases for tetanus in the United States continues at historically low levels [22 cases were reported to the Centers for Disease Control and Prevention (CDC) in 2006]. Tetanus disproportionately affects older Americans because many never received the series of tetanus vaccinations. For example, in 2005, two fatalities were attributed to tetanus: a 94-year-old woman who never received a tetanus vaccination and a 73-year-old woman with an unknown vaccination history. No neonatal cases were reported; in fact, no neonatal cases have been reported to the CDC since 2001.

QUESTIONS TO CONSIDER

1. What are some characteristics of the bacterial species that caused disease in Eduardo?

2. Summarize the treatments used to help Eduardo through this encounter, and indicate the reason for each.

3. List several ways the bacterial agent of tetanus might have entered Eduardo's tissues.

4. How do antitoxins and antibiotics each help to eliminate tetanus from the body?

5. In several places in this encounter, the writer refers to placing Eduardo in a quiet room and minimizing any outside noise. Why was that necessary?

A misdiagnosis can occur when physicians overlook the important information that a patient's story can provide. In this encounter, an emergency room physician uses his experience and intuition and, noting a single criterion, returns a diagnosis startlingly different from that of his colleagues. His diagnosis comes just in time to avert the patient's death.

A STRING OF PEARLS

TONY DAJER

Through the cubicle's curtains she looked like a middle-aged Doris Day, pink housedress buttoned to the neck and every blond hair sprayed in place. But her posture spoiled the effect. She sat cross-legged on the emergency-room stretcher, bent forward like an ancient pagan at morning ablutions. As I watched, she placidly, almost dutifully, delivered into the plastic basin cradled in her lap a stream of clear yellow vomit.

I looked around for Terry to get a bit of the story. Terry was one of the best residents—low-key, smart, and very conscientious.

"Hey, thanks for letting me grab lunch," I said. "Anything up?"

"Not much. Just this patient of Dr. Morgan's. I saw her quickly, but he had already called Dr. Summers, the gastroenterology consult, so I sort of backed off."

Terry sounded apologetic, but she shouldn't have. Morgan was the chief of medicine, and the lines of responsibility blur a bit when private patients show up in the emergency room. This was the boss's patient. The fewer cooks in the kitchen, the better.

"What's her story?" I asked.

"Her name is Mrs. Fratelli. She's 59 years old, with a history of insulin-dependent diabetes. Says she's been throwing up for two days, since she ate some bad cottage cheese. Hasn't been able to keep anything down, but swears she hasn't missed any insulin doses. No fever, no diarrhea, no headache, no chest pain. No nothing, really. Just vomiting. Summers wants to admit her and give her intravenous fluids. He thinks it's just garden-variety food poisoning. Morgan is on his way down to see her, too."

"Get any lab tests back yet?"

"Just her finger-stick glucose. It's 415."

The 415 thudded like a speed bump. Though normal glucose levels shouldn't go much over 100, diabetics can hit levels as high as 200 or even 300, especially when they're sick. But 415 was too high for garden-variety anything. I decided to take a look.

Mrs. Fratelli still had the plastic basin perched on her knees. Her look of calm concentration hadn't wavered.

"Hello, Mrs. Fratelli," I said. "I'm Dr. Dajer. I'm in charge of the emergency room."

She nodded, her eyes fixed on her offering bowl. I crossed my arms on top of her stretcher railing and considered the evidence. So far, my clues were high glucose and a plastic emesis basin. Only one slightly ridiculous question occurred to me.

"Mrs. Fratelli, do you feel sick?"

Mercifully, she didn't roll her eyes or guffaw. Instead she thought for a moment, patted her belly gently, and replied, "Well, I've got the heaves, of course. And I feel so very tired. But nothing else, I think."

"Having any trouble urinating?"

"No, not really. Maybe a little more than usual."

The rest of her answers were equally unrevealing. I glanced at her chart: Temp 98.9, normal. Pulse 88, normal. Respiratory rate 18, a touch high. Blood pressure 140/60, diastolic of 60 a touch low. No help there.

When patients can't point and say where it hurts, doctors' brains can skid like ten-wheelers hitting a slick curve. The diagnostic refuge of the perplexed becomes "viral syndrome" or "food poisoning," which doctors are fond of calling "gastroenteritis."

I rearranged my clues. Unexplained high glucose in a middle-aged diabetic woman equals—?

"Equals urinary infection, dummy," said a little voice in my head.

Doctors pay too much tuition to be taught rules of thumb; instead we are bequeathed clinical "pearls." And this pearl's corollary gave me another clue. Because diabetics can suffer nerve damage from their high glucose levels, they may not experience the discomfort that is the most common sign of bladder infection.

"Mrs. Fratelli, I know the nurses have already taken a urine sample, but would you mind giving me another one? There's something I'd like to check right away. Maybe we can start making you feel a little better, quicker."

She smiled for the first time. "Can't argue with that now, can I?"

Five minutes later the little dipstick that detects white cells in the urine was turning a dark purple. No doubt about it, Mrs. Fratelli had a urinary infection. A second pearl slid into place: simple bladder infections don't make you throw up. I spied Barb, the nurse taking care of Mrs. Fratelli, across the ER. She was rushing somewhere, holding two IV bags.

"Barb, hold up," I half-shouted, walking quickly after her. "Quick question: Mrs. Fratelli, the lady in 6A—is her temp oral or rectal?"

Barb thought for a second. "Oral, why?"

"I think it's more than food poisoning. She's got a definite bladder infection. Maybe more. Any chance of a rectal temp?"

Barb gave me a wry smile. I was notorious for requesting them. But a third pearl states: no temp is a real temp until it's a rectal temp. Sick patients breathe quickly, making the mouth too cool and dry to reflect the body's true temperature.

A bit later Barb was back. She said, "Okay, okay, just don't say I told you so. Her rectal temp is 101.7."

Clinical pearl number four clicked into place: infected urine plus fever plus vomiting equals sepsis.

As feared as it is deceptive, sepsis takes its name from the Greek word meaning "to putrefy." Known for generations as "blood poisoning," it generally means that bacterial invaders have managed to breach the natural barriers of bladder, lung, and skin and enter the bloodstream. Once there, they begin spreading a witches' brew of toxins produced by the bacteria or formed from the bacterial cell wall. In either case, they trigger a devastating immune response, the biological equivalent of calling a napalm strike on your own position. The body's inflammatory proteins, called cytokines, cause blood vessels to collapse and leak, tissues to swell, lungs to swamp, and kidneys to fail. The result can be septic shock so severe that no amount of intravenous fluid or artery-constricting medication can bring the sagging blood pressure back up. Roughly half of all patients with septic shock die.

For such a tidal wave of a disease, it is astonishing how subtle the first ripple of trouble can be. Even worse, those at greatest risk—older patients—are the toughest to diagnose. They may, like Mrs. Fratelli, come in because they are vomiting. Or they may feel confused, short of breath, or somehow just not right. Their complaints often sound nothing like the "classic" sepsis doctors are trained to recognize. Because we spend so much of our residencies caring for patients in the intensive care unit, doctors tend to become acquainted with sepsis only in the gravely ill. These patients spend weeks and even months with ventilators and multiple intravenous lines providing direct access for invasive bacteria. It is easy to forget that sepsis can afflict patients who walk in off the street and seem okay until, quite literally, they crash against death's door.

To make matters worse, doctors have been hamstrung by a clinical imperative: avoid bacterial overkill. Before blasting an infection with antibiotics, we are supposed to pinpoint the source with surgical precision. To do otherwise exposes bacteria to antibiotics needlessly, increasing the chance that they may develop resistance to the drugs we depend on to destroy them. So if we don't immediately find, say, pneumonia or an infection in the urine or skin, we are tempted to wait until a patient's illness declares itself through a lab test. But the only test that proves sepsis is a blood culture, which takes two days to turn positive. By then, the only declaration would be on the patient's death certificate.

Though it may seem appallingly obvious that one should treat blood infections posthaste, some early evidence misleadingly suggested that antibiotics were ineffective. It wasn't until 1980 that a landmark clinical trial clearly showed that giving antibiotics—and the sooner the better—cut deaths from sepsis in half. But medical thinking changes slowly; the new hurry-up philosophy has not yet reached the mainstream.

Moreover, with the rise of biotechnology over the past decade, researchers tried to design antibodies that could intercept bacterial toxins or the body's kamikaze cytokines, stopping the septic response in its tracks. Yet hundreds of millions of dollars in research and dozens of clinical trials have yielded only dismal results. Worse yet, dazzled by the prospect of "designer antibodies," doctors have neglected the basics. One large and very expensive study found that it took an average of 19 hours to start treatment once sepsis had been diagnosed. Emergency room doctors still routinely defer antibiotic orders to admitting doctors who, in turn, complacently wait for nurses to get around to it when they can.

The key, as the brilliant Confederate general Nathan Bedford Forrest put it, is "to get there first with the most men."

Mrs. Fratelli, I was convinced, had sepsis. I walked over to where Dr. Summers sat writing admission orders. Dr. Morgan stood at his side.

"Hello," I said. "What do you think about Mrs. Fratelli?"

"Oh, hi, Tony," Morgan replied. "Seems this cottage cheese gave her quite a turn. Probably gastroenteritis."

"Gastroenteritis," Summers nodded. "We weren't sure whether to cover her with antibiotics."

I tried to keep my voice casual. "Did you know she has a fever? Rectal temp's 101.7. And her urine's infected. I dipped it myself."

Summers was unmoved. "Well, she did eat that spoiled cottage cheese, you know. Been vomiting ever since. It's gastroenteritis."

My mind flashed back two months. A 60-year-old woman had come into the ER. Her symptoms were vomiting and low blood pressure. When flecks of blood showed up in her vomit, everyone figured her blood pressure was low because of bleeding from a stomach ulcer. A gastroenterologist examined her with a fiber-optic scope and saw only a few lesions in the stomach wall. Her oral temp read 98. No one took a rectal temp for six hours, but when they did, it was 104. She was vomiting up blood because a bladder infection had spread to her bloodstream, triggering sepsis. Vomiting in that weakened state had produced lesions in the stomach wall. Everyone had been fooled.

I came on duty after she'd already gone into cardiac arrest several times. We resuscitated her again and again, to no avail. Her husband, his face frozen with grief, kept asking me, "How could this happen? She was fine yesterday. We went for a long walk. How could this happen?"

Two days later her blood cultures grew *E. coli*, the bacterium that causes most bladder infections. The delay in starting antibiotics had squandered her only chance of survival.

"Gastroenteritis," Summers said again, as if repetition would mollify me. "What else could it be?"

"Look, I've seen this before," I said, trying to keep my voice even and authoritative. "She's vomiting because she's septic, and she's septic from her bladder infection. She needs amp and gent, now. Right away."

When used in combination, the antibiotics ampicillin and gentamicin kill the bacteria that are the likely cause of urinary infections. I was ready to order them myself. Private patient or no private patient, this was still my ER.

We argued back and forth for a while, with both Morgan and Summers gently trying to suggest I was crying wolf. True, the robust woman sitting up in 6A didn't look ready to succumb to a catastrophic illness. But I remembered that other woman going into repeated cardiac arrest. And I remembered her shattered husband. Finally, more to get me off their backs than because they were convinced, they gave in.

I headed down the hallway to Mrs. Fratelli's room, where I found Barb already at her bedside.

"Amp one gram and gent 120 milligrams stat," I told her. "Boy, one little rectal temp and you start calling in the heavy artillery," she tut-tutted.

I took up my position at Mrs. Fratelli's railing. "We think we know what's causing the trouble," I began. "If we're right, you might even be able to get rid of that basin soon."

My patient stared into her lap. "And just in time, too. You know, I just realized it doesn't go with my robe."

Two days later I met Morgan outside Mrs. Fratelli's room.

"How's she doing?" I asked.

"Fine."

"Seen the lab results?"

"Not yet. Anything growing in the blood cultures?"

I'd just been to the lab. "*E. coli*," I answered. It was clear now that Mrs. Fratelli was suffering from more than a case of food poisoning from bad cottage cheese.

Morgan's face brightened like a naturalist's before a rare and brilliantly colored butterfly.

"No! Good for you!" He smiled sheepishly, patted me on the back, and added jokingly, "You know, I just knew we should have taken a culture of that cottage cheese."

UPDATE

Sepsis, also called systemic inflammatory response syndrome (SIRS), involves symptoms often resulting, as this encounter described, from the body's inflammatory response to a bacterial infection, such as Escherichia coli, *or to a viral, fungal, or parasitic infection. Researchers predict that in 2007, one million Americans will develop sepsis and 30 percent will die, making it the 10th leading cause of death. SIRS is the leading cause of death in noncoronary intensive care unit (ICU) patients and is reported in 1 to 2 percent of all hospitalizations. It is a major cause of death in ICUs worldwide, where mortality rates range from 20 to 40 percent.*

QUESTIONS TO CONSIDER

1. What treatment might have been administered to Mrs. Fratelli if the original diagnosis of gastroenteritis had been upheld?

2. What symptoms would Mrs. Fratelli have experienced if she had not been diabetic? How did her diabetes mask the symptoms?

3. Why are white blood cells in the urine a sign of infection?

4. Why is taking a rectal temperature often more reliable than an oral temperature?

5. Why did the writer title this article "A String of Pearls?"

In most patients, brain seizures reflect a serious underlying condition. In this encounter, physicians discover that a deadly viral disease has been passed among members of a family and that denial of Western medical principles has led to sad consequences. Indeed, the family members have denied the very existence of the infectious agent.

A DEADLY SPECTER

JEREMY BROWN

After 12 hours in a busy emergency room, Peter, the intern I was relieving, looked exhausted. His busy day shift had ended and it was time for me to take over. As we walked past each patient's room, Peter delivered a one-sentence summary. "Room two, 22-year-old with appendicitis, surgery has seen her, OR booked. Room three, 85-year-old nursing home resident, fever for two days, chest films need to be checked, urine looks clear." Peter paused outside the room of his most recent arrival. "Room one, 32-year-old Haitian, took some witch doctor medicine, is complaining that he has no feeling in half of his body. Psychiatry has been called. Good night." And with that, Peter headed for home.

I walked in to evaluate Henry Pierre, the patient in room one. When I had arrived for my shift, the paramedics who brought Henry in had told me his story. He had been fine when he sat down to dinner with his girlfriend. But in the midst of the meal, he had become violent for no reason and she had called an ambulance. The EMS crew had matter-of-factly described finding Henry thrashing about on the floor, cursing and sweaty. It had taken two sturdy paramedics and two even sturdier Boston police officers to wrestle Henry off the floor and into the ambulance. Naturally, I was a little wary.

"Is Mr. Pierre restrained?" I asked a nurse.

"Oh yes," she replied. "Four-point restraint—good enough for you?"

I quickly calculated that only Houdini could escape from locked restraints on each limb faster than I could call for help. Comforted by this thought, I entered Henry's room.

Henry lay in bed, smiling and calm. He certainly didn't look violent, or even ill for that matter. I introduced myself and asked what had happened, fully expecting some strange story about people who were out to get him. His reply was even stranger. Henry had absolutely no recollection of his violent episode. He didn't even remember how he got to the hospital. In fact, the last thing he remembered was eating dinner with his girlfriend. The way Henry was smiling reminded me of someone who has been hypnotized and can't understand why his hand keeps rising every time the hypnotist says the magic word. He seemed to feel nervously amused at his situation. He knew where he was, and he answered all my questions as well as anyone might under those conditions.

"So you won't get violent again if I remove the restraints?" I asked.

Henry's smile grew even larger. "How could I?" he replied. "I can't move my right arm or leg."

When I tested his muscle strength, the left side of his body was normal, but there was no response—not even a twitch—in his right leg. And he could barely move his right arm. I decided to unfasten his restraints. After all, even if he did get violent, I reasoned that I could run faster on two legs than he could on one.

Henry told me that his right side had been getting weaker each day for the past week. In fact, he had had to stop driving his cab because his right leg was too weak to depress the accelerator. I then tested his ability to feel pain. When I pricked his left leg with a pin, he flinched. But when I tested his right side, I got no response. Though the pinpricks on his right leg had drawn a little blood, Henry couldn't feel them.

If Henry had been psychotic when he arrived, he was quite rational now, and although psychiatric patients do strange things, his lack of sensation was too convincing to be feigned. Of course, his violent episode at home needed to be explained, but right now, I decided, Henry didn't need a psychiatrist. He needed a radiologist.

I strongly suspected that something in Henry's brain was causing his bizarre behavior and lack of sensation. Only after we were sure there was nothing structurally wrong with Henry's brain could we pass his case on to the psychiatrists. When I called the psychiatrist covering the ER, she was happy to hear that she needn't come down to see Henry. The radiologist on call, however, wasn't too happy to hear that I needed an emergency CT scan of Henry's brain.

While we waited for the radiologist to arrive, I got to know Henry a little better. He had arrived from Haiti 16 years earlier, and he had been working as a cabdriver ever since. He said he had been perfectly healthy until a week before, so healthy he had never even visited a doctor. Even when he noticed that his right arm and leg were weakening over the past few days, Henry had turned not to conventional medicine but to a voodoo belief system that I poorly understood. Growing up in Haiti, where African healing traditions are strong and trust in Western medicine is weak, Henry had always consulted a bocor when he was feeling ill. The bocor (the word is Creole for voodoo healer) would give him some medicine, and he would get better. Just the day before, Henry's brother, who was a full-time bocor in a large Haitian community in New Jersey, had driven to Boston to treat Henry with some herbs he had mixed into wine.

I would later learn that it was this use of alternative medicine, together with Henry's bizarre loss of feeling, that had prompted Peter to refer Henry to a psychiatrist. Many doctors, trained in a Western medical tradition in which scientific explanation reigns supreme, would have done the same. We find it difficult to accept other ways of understanding the world, and this difficulty is proportional to our ignorance. If Henry had merely prayed to a Judeo-Christian God or gone to confession, few physicians would have called the psychiatrist. But to many doctors, taking voodoo medicine for paralysis sounds crazy. Even in the multicultural climate of the nineties, such reactions are more common than we may care to imagine. Peter had assumed that Henry's paralysis was a hysterical reaction of some sort and therefore assumed that a psychiatrist would be the best person to diagnose and treat the condition. It was an understandable, but nevertheless mistaken, assumption.

Henry's CT scan showed two large masses—each over an inch square—on the left side of his brain. His condition was extremely serious. Such masses are caused by abscesses, tumors, or infections, and as they grow larger, they can produce an array of neurological symptoms. According to the radiologist, Henry's CT results clearly pointed toward infection. When the brain is fighting an infection, the CT scans show dying tissue surrounded by a bright ring of inflamed living tissue crammed with immune cells battling the infection. The most likely cause of Henry's infection was an intracellular parasite called *Toxoplasma gondii.*

More troubling than the infection itself, however, was that *T. gondii* is most likely to cause disease among people with a severely crippled immune system. People tend to get infected with the parasite through contact with contaminated cat feces or raw meat, but if their immune systems are strong enough to limit the spread of infection, they don't show any signs of disease. The exception to this rule is the fetus. If a woman becomes infected while she is pregnant, the parasite can infect the fetus through the placenta, sometimes leading to severe complications and even death.

Henry had already told me he wasn't taking any kind of medication or drug that would weaken his immune system. That meant some other pathogen was wiping out his immune system. In the vast majority of toxoplasmosis cases, that pathogen is HIV. Once HIV has overwhelmed the normal immune mechanisms that prevent the parasite from spreading, *T. gondii* tends to attack the brain. The infected brain tissue attracts immune cells, and the resulting inflammation and swelling can cause seizures and loss of sensation.

It was rare—but not impossible—for HIV to announce itself this way. But I was puzzled. From what Henry had told me, he had none of the most common risk factors for HIV infection. He didn't use intravenous drugs, and he hadn't had unprotected sex with an infected partner. Moreover, the symptoms of HIV infection tend to follow a common pattern. When a person first becomes infected, there is usually a mild flulike illness, followed some months or years later by the symptoms of AIDS—night sweats, weight loss, sinusitis, diarrhea, and coughing. Yet Henry told me he had been perfectly healthy until the numbness set in two days ago. Now I was assuming he had a massive brain infection because of suspected HIV infection. I had to find out if my suspicion was correct.

When I got to Henry's room, I found him surrounded by several family members. Although I was uncomfortable talking about the CT scan results in their presence, no cajoling on my part could get them out of Henry's small room.

"Henry, the tests show us that you need to come into the hospital right away," I said carefully. "You have something in your brain that may have caused both the weakness and your violent episode today."

Was this how you were supposed to break the news? Was I being sympathetic or totally missing the mark? I glanced over my shoulder at Henry's relatives.

"What does he have in his head?" his sister asked.

Given the little information I had, I replied in the most general of terms. "Well, it may be an abscess, or a growth like a tumor, or an infection."

Just then, Henry's eyes began rolling to one side and his body began shaking violently. He was having another seizure. This time, instead of being restrained by lock and key, Henry was pinned down by his relatives, who were each hanging onto a limb and yelling in Creole. I tried to keep everybody calm, meanwhile frantically struggling to keep Henry's arm still enough to inject an antiepileptic drug and a quick-acting tranquilizer. Needle finally hit vein and Henry dropped off into a deep sleep. He was wheeled out of the ER and out of my care. When he would awake several hours later, he would be in the care of a neurologist.

Though emergency room doctors are often gratified to be the first to diagnose and treat a disease, our contact with patients is often frustratingly brief. When I finally got a chance to look in on Henry in the neurology ward, he was smiling, just as he had been when we first met. Treatment with an antibiotic had cleared up his brain infection, and he had quickly regained the use of his right side. But my suspicion had been confirmed. Henry had taken the test for HIV; and the result was positive.

I thought back to my talk with Henry in the ER. I had somewhat naively thought that he would reveal to a complete stranger the intimate details of his personal life. Many of us have secrets that we find difficult to admit even to ourselves, let alone to others. The coping mechanisms that we use to deal with our problems are as complicated as the conditions they help us handle. While the ways people cope with the prospect of dying from a fatal disease have been well analyzed by psychiatrists, psychologists, social workers, and chaplains, the truth is that nobody ever plays it by the book. Most patients with a fatal disease deny it at first, then gradually come to accept their condition. But Henry, perhaps because of his underlying resistance to Western medicine, or perhaps because of the sheer psychological burden of living with the virus circulating within him, had never gotten past the initial stage of denial. When I later learned the painful truth about Henry and his disease, his reluctance to confide in me became more understandable.

After a few weeks in the hospital, Henry had finally told one of his doctors that his girlfriend was HIV positive. He said he had known this for many years but had chosen to ignore the risks that unprotected sexual relations with her would bring. Rather than accept the unacceptable, he had chosen to deny the very existence of the disease, and he lived his life as if nothing were wrong. But the existence of the virus was something no denying would ever change.

Perhaps the saddest part of Henry's story was that he had already lost a two-year-old son to the disease. A year before, the toddler had died at another hospital of an overwhelming chest infection. A quick phone call to that hospital by Henry's physician confirmed that the boy had died of *Pneumocystis carinii* pneumonia, a common and often fatal complication of HIV infection. The child had become infected with HIV while still in his mother's womb. Although Henry acknowledged that his son had died, he claimed that the boy had died of bad pneumonia. The child had indeed died of pneumonia, but it was pneumonia caused by HIV infection. Henry's denial was not limited to his own infection.

Henry spent three weeks in the hospital, and when he left, he had regained the complete use of his right arm and leg. He was able to drive his taxi again, which allowed him to provide for his girlfriend and their two other children. He reluctantly agreed to allow a hospital social worker to work with the family to ensure that the parents both stayed healthy for as long as possible.

A few months later I ran into Henry on his way to a checkup with his doctor. He said he was planning a trip back to Haiti for a vacation and a visit to the best bocor he could afford. That, he assured me, would eliminate the bad spirit— he never referred to HIV—and he hoped to return to the United States happy and healthy. Indeed Henry looked happy and healthy. He was still smiling, and he showed none of the common symptoms of AIDS.

As he walked away, I reminded myself of how deceptive appearances can be. Henry and his girlfriend had chosen to deny her illness, with tragic consequences. I know that denial can be a normal phase of coping with illness, and I sometimes wonder how I would handle living with the specter of HIV infection. Even after his seizures, hospitalization, and recovery, Henry Pierre clung to his traditional views. Some doctors might say that Henry needed to see a psychiatrist after all.

UPDATE

For almost 30 years, the human immunodeficiency viruses (HIV) that cause AIDS (acquired immunodeficiency syndrome) have been spreading globally. The Centers for Disease Control and Prevention (CDC) estimate that in 2007 about 40,000 Americans have become infected with HIV and more than a million Americans are living with HIV. However, the disturbing news is that some 25 percent of HIV-infected Americans don't know it! So the message is clear—if you don't know your HIV status, get tested. There are several new antiretroviral drugs available today and better therapies, so HIV-positive individuals can live better and longer lives; in fact, the drugs and therapies can delay the progression from HIV disease to AIDS. But the first step is to get tested.

QUESTIONS TO CONSIDER

1. How might Henry have acquired the infectious agent responsible for the brain seizures?

2. How did the CT scan indicate an infection in the brain?

3. What steps might Henry have taken to prevent contracting the virus from his girlfriend?

4. The physician referred Henry to a social worker. What will the social worker discuss with him?

5. As noted in this encounter, it is unusual for HIV to announce itself in this way. What are the normal symptoms of an HIV infection?

The patient's severely depressed red blood cell count had the emergency room staff puzzled, but then one physician noticed the mahogany-colored urine. The encounter pointed to the most predominant infectious disease in the modern world and a key public health problem in the global community.

BLACKWATER FEVER

TONY DAJER

"He looks sick to me. Really sick."

The worried tone in the resident's voice, even amid the hubbub of a busy emergency room, caught my attention like a shout. I quickly finished with the chart I was writing and made myself available.

"What's the story?" I asked her.

"Twenty-year-old Chinese male came in last night complaining of headaches and fevers. The night team thought it might be meningitis, so they did a spinal tap. But the lab just called, and the fluid's clear, so there's no meningitis. And now his temperature is back up to 103."

Before Barbara, the resident, had found me, she had mentioned the case to the staff neurologist, who was in the emergency room to examine someone else. A quick detour into our patient's room was enough to plant a worried frown on his owlish features.

"I can't find anything abnormal," he said, after giving our patient a quick neurological exam. There was nothing that might indicate a brain tumor, stroke, abscess, or other obvious source of head pain. "Unfortunately he doesn't speak any English, so it's tough to get the story. But one thing's for sure:

he's got a heck of a headache." He suggested it might not be a bad idea to get a CT scan.

As we clustered outside the patient's room, a nurse handed us the results of his second blood count. We all gaped. Six hours earlier the fraction of his blood made up of red cells had been 36 percent—low, but not dangerous. Now it was 30 percent. Right under our noses he had lost the number of red cells contained in almost a quart of blood—and nobody had the faintest clue why.

I turned to Barbara. "Where's this kid from?"

"Southern China, we think," came the reply. "There were no interpreters who spoke Cantonese in the hospital last night. Two of the night team examined him, and as best as they could tell, he's been in the States for three or four months. Apparently he's had bad headaches for a while."

As if unleashed from deep within some mnemonic curio box, a swarm of tropical diseases suddenly came to mind, a veritable menagerie of illnesses caused by protozoa, nematodes, fungi, viruses, and bacteria. But no obvious candidates to explain the young man's symptoms came into focus.

"Okay, before we do anything else, let me take a look," I said, more to fulfill my role as Barbara's supervisor than out of any hope, as the fifth doctor in line, of sniffing out any missing clues.

After a parting "Good luck" from the neurologist, we trooped back into the exam room. I gave the young patient my best friendly wave, but all he could do was hold his head in a way that says "agony" in any language. I felt his forehead. It was burning. His sheets were soaked. But with the rest of his exam I drew a blank. Everything was normal: neck supple, lungs clear, heart regular, abdomen soft, skin unblemished. When I was done, Barbara and I quickly reviewed where we were and what we knew. The answer was, to quote Dashiell Hammett: "Nowhere and nothing."

There was only one thing to do. I tilted my head back and shouted: "Anybody here speak Cantonese?" Two nurses turned and pointed to Judith.

Judith was one of the best nurses in the emergency room—indispensable for keeping young residents out of trouble and older attending physicians in line. I trotted over to her and clasped my hands together in a gesture of supplication.

With Barbara and me in tow, Judith approached the young man's bedside. I told her what little we knew of his medical history, then asked her to take it from the top: How long had he been in the States? When did he get sick? From the stream of rapid-fire Cantonese that followed, I realized she was, unprompted, moving on to the next several questions on my list.

While we waited for Judith's translation, I glanced at the patient's clear plastic urinal. It was almost full and the urine was brown, a deep mahogany brown. Suddenly, as if it had been kicked in just the right spot, my brain's gearbox came unstuck.

"Have you ever heard the term blackwater fever?" I asked Barbara.

"Um, no," she answered, then gave me an expectant look.

I pointed to the urinal.

In the old days they used to call falciparum malaria—the most severe form of the disease—blackwater fever because the intense destruction of red cells it causes turns a patient's urine black. My hunch seemed farfetched, though. Many other diseases cause dark urine, and malaria is no longer indigenous to this country. Virtually all U.S. cases—about 1,100 in 1990—are found in travelers or newly arrived immigrants. Most American doctors never even see a case. I could hardly wait to hear where this guy had been.

Judith supplied us with the answers. "He's only been in the States three weeks, not three months. He's had fevers and headaches since he got off the plane. Went to three doctors in Chinatown who gave him pills, but none helped."

"Where did he fly in from?" I asked.

"Thailand," said Judith.

"And where did he start out from?"

"Southern China."

But, I dimly remembered, Thailand and China had no common border. "How did he get to Thailand?"

"He left China by crossing through Burma," came the reply. "On foot. All the way. The journey took about two weeks."

Now his symptoms began to make more sense. Eastern Burma is a hotbed of malaria, especially falciparum malaria, which is resistant to chloroquine, the standard drug used to treat the disease. And the incubation period for malaria is two weeks. Had he been bitten by a malaria-carrying mosquito in the jungles of Burma, the fevers and headaches wouldn't have started until he arrived in the United States.

For a firm diagnosis, however, a smear of blood would have to be put under the microscope and inspected for malaria parasites. Bloodsucking mosquitoes merely transmit malaria from one human host to the next; the actual cause of the disease is one of several tiny parasites that attack the red cells in human blood. The exam would tell us which strain of malaria the young man had—provided, of course, that it was malaria we were dealing with. We sent off a tube of blood to the lab right away, but it would take 30 minutes to prepare the slide. Not that we had time to sit around; there were plenty of other patients in the emergency room with maladies less exotic but no less urgent.

Malaria is now rare in the United States, but it once was a killer. It loosened its grip on the American South only in the late 1940s. Over the centuries it has brought down empires, decimated armies, and depopulated vast tracts of land, especially in equatorial countries. Indeed, the force of its onslaught is inscribed in our very genes. Sickle-cell anemia is a major health problem among African Americans; it arises from a mutation in the gene for hemoglobin, which carries oxygen and colors blood cells red. Before the advent of modern medical care, inheriting a pair of sickle genes, one from each parent, would have doomed a person to a painful early death. But inheriting only one sickle gene offered a critical advantage: greater resistance to malaria. In other words, a potentially deadly gene flourished because, for some, it conferred protection against malaria. Unfortunately, for an individual living in a malaria-free country such as ours, the mutant gene no longer has benefits, only costs.

In the tropical world, however, malaria remains a plague. In 1991 some 150 million people contracted it, and almost 2 million—mostly children—died. Malaria has not only withstood everything that modern medicine has thrown at it but it is even regaining lost ground. In the 1950s global campaigns were launched to stamp out the disease and its pesky mosquito vector. DDT became the new miracle insecticide, and chloroquine the new antimalaria "magic bullet." Yet 35 years and billions of dollars later, all that has changed is that from Peru to Nigeria to Vietnam, mosquitoes have become DDT-resistant and falciparum malaria has become impervious to chloroquine. Meanwhile, even the world's greatest immunology labs have failed to come up with a vaccine that works.

As a final slap in the face to modern technology, the ultimate drug against malaria is still quinine, just as it was 300 years ago. And old prevention methods, such as using mosquito nets and draining stagnant pools where mosquitoes breed, have recaptured center stage. Our ancient foe, it seems, is sending us a very modern warning about nature's capacity for retaliatory strikes.

Here, though, I wasn't going to let this grim disease gain the upper hand. But before we could treat our patient, we still needed confirmation of my diagnostic hunch. Finally, we got the call from the lab. Barbara grabbed the phone. It took only seconds to relay the message.

"It's malaria!" she cried. "And it looks like falciparum."

I collared the other resident on duty.

"Come on, you won't see a slide like this every day."

The three of us charged out of the emergency room and grabbed the elevator to the lab. The technician had the slide ready. She adjusted the microscope.

"There, right in the middle of the field, you can see the 'signet ring.' "

My eyes adjusted to the scope, and I focused on several of the pale pink blood cells. Smack in the middle of one of them was the bluish circular band crowned by a tiny ruby: the one-celled protozoan named *Plasmodium falciparum*. Yet there was no obvious evidence of the destruction wrought by this pretty killer. That, too, I learned, is one of falciparum's tricks. It doesn't simply do its dirty work by destroying red cells. It also makes their membranes sticky and causes them to clump in vital organs such as the liver, the kidneys, and the brain, cutting off their blood flow. The result is often permanent organ damage or, in the case of cerebral malaria, death.

The residents took turns looking through the microscope. Exclamations filled the room as an image they had seen only in textbooks came into focus. We had our diagnosis. Now it was just a matter of administering the right dose of quinine, adding a boost of tetracycline, and our young visitor would be cured.

On our way out of the lab, we looked like a team trading high fives in the end zone. As I looked at the happy faces, I knew mine was no different. It occurred to me then what an odd bunch we doctors are—celebrating tiny smudges on microscope slides and all articulating the same astonished monosyllable in the face of a deadly pathogen: "Wow!"

UPDATE

The number of reported cases of malaria in the United States has remained relatively stable for the past 15 years. Nearly all of these infections have occurred in persons who traveled to or emigrated from a malaria-endemic country, such as this encounter described [in 2005, the Centers for Disease Control and Prevention (CDC) reported almost 1,500 such cases]. However, worldwide malaria is one of the most severe public health problems and is a leading cause of death and disease in many developing nations, with young children and pregnant women being primarily affected. According to the World Health Organization (WHO), in 2007 more than 3 billion people (half the world's population) live in areas at risk of malaria transmission. The WHO reports that there are 350 million to 500 million clinical episodes of malaria and more than one million deaths every year due to malaria. About 60 percent of the malaria cases worldwide and some 80 percent of malaria deaths worldwide occur in sub-Saharan Africa.

QUESTIONS TO CONSIDER

1. Name two reasons why the disease in this encounter has been brought under control in the United States.

2. How is a serious drop in red blood cells related to this disease, and what is the key diagnostic information offered by the mahogany-color urine?

3. What is the value of maintaining a sickle cell gene in a human population where the disease in this encounter is widespread?

4. If the disease described in this encounter is so prevalent, why do you think there is currently no vaccine to prevent it?

5. The laboratory obtained valuable information after examining the patient's blood. What did the laboratory technologist and doctors observe?

Eating whole, natural foods can contribute to a healthy way of life, but as this encounter shows, there are certain hazards involved, especially when feeding natural foods to children. Indeed, the source of the infectious agent in this story is widely recognized, and routine warnings are issued to parents of infants and very young children.

THE BABY WHO STOPPED EATING

ROBERT MARION

"I saved the most interesting case for last," said Molly Wilson, the resident who'd been on call the night before. It was a Saturday morning in early February. Molly and I had spent the last hour touring the Infants' Unit with the interns, stopping to discuss and examine each child who was unlucky enough to be inhabiting the unit that day. I was tired, the day was cold and gray outside, and I'd much rather have been at home in bed. But as the attending physician that month, the most senior doctor on the service, it was my job to make sure these children got the best care possible, and so fighting off the urge to daydream, I focused my attention on the resident. "This baby's name is Jarret Fox," Molly continued. "He's a three-month-old who was admitted last night for dehydration. According to his mother, Jarret stopped eating four days ago."

"Stopped eating?" I repeated, quickly coming to full attention. "What do you mean he stopped eating?"

"Just that," Molly replied. "His mom says that Jarret was happy and healthy a week ago. Then, on Tuesday, he seemed to lose interest in nursing. He just stopped sucking, his mother says, and he hasn't eaten anything since."

"That can't be right," I responded. "Three-month-olds don't just suddenly stop nursing and starve themselves until they get dehydrated."

"Well, I didn't believe it either at first, but the mother keeps telling the same story: she's been trying to force-feed him since Wednesday but hasn't had any success. Yesterday she brought him to her pediatrician. He said Jarret was about 5 percent dehydrated. He also said the kid was much floppier than he'd been the last time he'd seen him. So he sent him in for rehydration and a full evaluation."

That last part of Molly's report, the part about the increased floppiness, made my heart sink. It suggested a condition I hoped this baby didn't have. "Do you have any ideas about a diagnosis?" I asked.

"The only thing I can think of is spinal muscular atrophy," Molly said.

"That's what I'm thinking, too," I replied. "I hope we're wrong. Let's go see him."

Put simply, a diagnosis of the infantile form of SMA is a death sentence. A relatively rare inherited disease in which the nerves that control movement mysteriously degenerate and disappear, it is the childhood equivalent of the better-known (but no better understood) amyotrophic lateral sclerosis. As the nerves vanish during the first months of life, a child with SMA grows progressively weaker. After initial problems with feeding, the infant loses the ability to move its arms and legs. Breathing also becomes difficult. With time, the child becomes more and more hungry for air until finally, by about the first birthday, he or she dies. The cause is usually pneumonia, a common infection in lungs that aren't getting enough air.

As a medical geneticist, I have had the unenviable task of helplessly watching more than a dozen patients live out the nightmarish symptoms of SMA. The only thing I could do was aid families in coping with the loss of their children. As I entered Jarret Fox's hospital room that Saturday morning, the faces of all these children and their families flashed through my mind.

"Ms. Fox," Molly said as we approached Jarret's crib, "this is Dr. Marion. He's our attending pediatrician."

"Sorry we have to meet under these circumstances," I said with a smile as I shook her hand. Barefoot, clad in a peasant blouse and bell-bottom jeans, her long, straight hair parted down the middle, Jarret's mom looked like a long-lost refugee from the Summer of Love. She also looked as if she could use a good night's sleep. "How are you doing?"

"Not too well," she replied. "I'm hoping someone will be able to tell me what's wrong with my son."

"We're going to try to get to the bottom of it," I said. "First, maybe you can tell me the story from the beginning."

Without hesitation, Ms. Fox spilled out the short tale of her son's life. After an uncomplicated pregnancy, Jarret had been born at his parents' home in North Salem, a rural town north of New York City. He was the couple's second child: their daughter, Jessica, now three years old, was "healthy as a horse." Although his birth was attended only by a midwife, Jarret was examined on the first day of life by the family's pediatrician (the only one in the area who practiced homeopathic medicine and made house calls) and declared to be in excellent health. His mother could think of nothing unusual about her son's newborn period: in her words, he had been "like my other baby."

The infant had been seen by the pediatrician on a regular schedule, first at two weeks, then at a month, then at two months. He'd received his immunizations and had been growing and developing normally. Ms. Fox explained that her family were strict vegetarians who ate only whole, natural foods. She assured me that Jarret had had nothing but breast milk, adding proudly, "My daughter was exclusively breast-fed for the first 18 months of her life."

But four days ago this idyllic existence had ended. Jarret had simply refused to nurse. "He just wouldn't latch onto my breast," she said sadly. "Nothing I did got him interested. It was like a switch had been turned off in his brain and he wouldn't do it anymore. Just like that."

"Has he been hungry?" I asked, less certain now about the diagnosis.

"At first he was," she said. "That first day, he cried and cried. It was pathetic. But since then, he's just been lifeless, like he just doesn't care anymore."

I could see what Ms. Fox meant. Jarret was a sturdy, beautiful baby, but he lay as limp as a rag doll in his hospital crib, an IV in his left arm and a feeding tube in his left nostril. Although his eyes returned my gaze, Jarret seemed passive and expressionless.

"This doesn't sound like SMA," I said, shaking my head. After finishing my examination, I thanked Ms. Fox and told her that we needed to speak with the neurologist and that we'd be back later. Molly, a few interns, and I assembled in the corridor.

"SMA doesn't start suddenly like this," I began. "The weakness comes on gradually—the first day, the parents notice that the kid's a little floppy, the next day he's a bit more floppy, then a little more floppy the next, until finally they find they can't get him to eat enough to keep himself going. That's when the kid comes to the hospital with dehydration and the diagnosis is made. But this story of the weakness coming on suddenly like a switch going off—that's too acute to be SMA!"

"I agree," Molly said. "It sounds almost like the kid was poisoned."

"Poisoned by what?" one of the interns asked. "The kid has had nothing but breast milk. If he was poisoned by something in the breast milk, the mother should have been affected, too."

"Good point," I replied, as a little bell of recognition began ringing in my head. "But Molly's right. It does sound as if he's been poisoned. And I think I know what it was." Without another word, I headed back into Jarret's room with the rest of the ward team trailing behind.

The mother, who had been sitting beside Jarret's hospital crib, rose to her feet.

"Sorry to bother you," I said. "But tell me again, when did you first notice this change in Jarret?"

"Tuesday afternoon," she replied. "When he woke up from his nap. He's usually starving when he wakes up. But that day, I couldn't get him to take my breast for anything."

I nodded. "And your three-year-old. Tell me, how does she get along with Jarret? Does she help you take care of him?"

"Oh, she's crazy about him," Ms. Fox replied with a smile. "She helps change his diapers, and when he spits up, she wipes him with a cloth. She tells me that I'm *her* mother, and she's really Jarret's mother."

I smiled at this also. "Since Jarret's exclusively breast-fed, she hasn't ever fed him, has she?"

"No, we'd never let her. But she always pretends to feed him. She pretends to spoon food into his mouth. It's really cute and they both love it."

"But as far as you know, she's never actually fed him?"

"Definitely not," Ms. Fox replied. "My husband and I are always at the table supervising. We'd never let Jess put anything in the baby's mouth."

I nodded and continued: "Ms. Fox, what does Jessica have for breakfast?"

The mother, somewhat surprised by the non sequitur, answered without hesitation: "A bowl of hot oatmeal and a glass of milk. Why do you ask?"

"Does Jessica eat the oatmeal plain, or does she put sugar on it?" I asked, already knowing what the answer would be.

As expected, Ms. Fox gave me an angry look. "Dr. Marion, we eat only whole, natural foods—no meat, no processed food, no sugar. Sugar is poison."

"Okay, no sugar," I pushed on. "But does Jessica use anything to sweeten her oatmeal?"

"We allow her to use a teaspoon or two of honey," she replied.

"Ms. Fox, we have to do some tests, but I think Jarret's going to be okay. I'm pretty sure he's got botulism."

It was Ms. Fox's reverence for natural foods that tipped me off to the possibility of infant botulism. That, and the suddenness of Jarret's symptoms. While considering the diagnosis as I'd questioned her, I visualized the scenario that had undoubtedly led to the baby's sudden onset of weakness.

Early that Tuesday morning, the Foxes were all in the kitchen. Jarret, sitting happily in his infant seat, had been placed at the table next to his sister, who was enjoying a bowl of oatmeal that had been topped with a few dollops of natural honey, straight from the hive. The children's parents had perhaps stepped away from the table to prepare their own breakfasts. Suddenly, Jessica, pretending to be Jarret's mother, silently offered her brother a spoonful of cereal. The infant eagerly accepted the offer and carefully rolled the strange-textured substance around in his mouth before swallowing. He smiled with satisfaction as Jessica, still in silence, finished off the bowl.

Later in the day, Jarret took his usual afternoon nap. When he awoke, his mother found that, mysteriously, he could no longer take her breast.

As Ms. Fox continued to answer my questions, I became more convinced that this scenario (or one like it) had occurred. It had to have: after hearing the story and seeing Jarret, there was no other logical explanation.

Like Ms. Fox, most Americans believe that when applied to foods, terms like "pure" and "natural" are synonymous with "healthy" and "nutritious." This may be accurate for most foods, but not honey. Eating honey—in both natural and processed form—can lead to serious disease or even death in infants. Because of the environment in which it's produced, unprocessed honey often contains spores of *Clostridium botulinum*, the bacterium that causes botulism. The same can be true for processed honey. In most humans, the spores cause no problems: immune cells in the intestinal tracts of older children and adults release proteins that readily bind to and destroy the toxin. But in children under one year of age, infants whose intestinal tracts are still immature, the *C. botulinum* survives and starts making toxins. And that spells big trouble. After traveling through the gut's lining and entering the bloodstream, the toxins are carried throughout the body, where they bind to peripheral motor nerves, preventing them from carrying messages from the central nervous system to the muscles. Within hours of ingesting even tiny amounts of contaminated honey, previously healthy infants become profoundly floppy and lethargic, unable to smile or cry or suck. If the dose of toxins is large enough, every muscle, including those involved in breathing, becomes paralyzed. If their condition is not recognized quickly, these infants may simply stop breathing and die.

But if the diagnosis is made early, the prognosis for full recovery is good. Although there is no antidote to the toxin, its grip on the nervous system weakens with time. Gradually, the motor neurons create new receptors to replace those blocked by the toxin.

If the child is supported through this period—if he is tube-fed, provided with oxygen, and placed on a ventilator if breathing becomes difficult—he will eventually return to the state he was in prior to the disorder. The period of paralysis can last weeks or months.

When I told Ms. Fox that I believed Jarret had botulism, she looked at me as if I was crazy. But when the neurologist came by a few minutes later and agreed with the diagnosis, she began to have second thoughts about her initial impression. Later, when an emergency electromyogram (a test of

Jarret's muscle and nerve function) revealed abnormal nerve responses consistent with botulism, she, too, became positively convinced of the story I'd invented.

Although we waited three long weeks for the lab reports, the results confirmed the presence of *C. botulinum* toxin not only in Jarret's serum and feces but in a specimen taken from the jar of honey from the Foxes' pantry as well. Because the scenario now seemed so obvious, I urged the Foxes not to confront or blame Jessica; doing so, I argued, would needlessly make the girl feel guilty. Rather, I suggested they have a talk with her, trying to get her to understand that she should never put anything into her little brother's mouth.

As for Jarret, it took him more than five weeks to return to his preposoned state, and his recovery was not without complication. On the afternoon of his admission to the hospital, his breathing had become labored, and when a blood-gas analysis revealed signs of respiratory failure, he was transferred to the ICU, where he was intubated and placed on a ventilator. For weeks he remained dependent on machines, unable to breathe, suck or swallow, cry or smile, or move any of his muscles. He continued to be fed milk pumped from his mother's breast (she wouldn't allow him to be fed anything else) through the feeding tube.

Then in early March, his nurse noted what appeared to be a flicker of movement in his left leg. It was so subtle at first that she thought she'd only imagined it, but more movement occurred in the following hours. Slowly but surely, Jarret was regaining control of his nervous system.

In the next few days, he was gradually weaned off the ventilator. Soon the feeding tube was removed, and he began eating on his own again, first from a syringe, then from a bottle, and finally, more than a month after he had entered the hospital, directly from his mother's breast. Just about back to his old self, he was discharged in the middle of March.

UPDATE

Besides infant botulism described in this encounter, there are two other types of botulism. Foodborne botulism results from the ingestion of foods containing the botulism neurotoxin, while wound botulism can result from a wound infected with Clostridium botulinum *and the neurotoxin it produces. All forms of botulism can be fatal and are considered medical emergencies. In 2006, the Centers for Disease Control and Prevention (CDC) reported 144 cases of botulism; about 10 percent were foodborne botulism, 58 percent were infant botulism, and 32 percent were wound botulism (or unspecified). Today, a purified and diluted neurotoxin called Botox® is administered in a medical setting as an injection for clinical and cosmetic use.*

QUESTIONS TO CONSIDER

1. Why is the ingestion of spores in infant botulism so dangerous to infants but not to older children and adults?

2. The disease described in this baby is generally related to foods packaged and stored a certain way. How?

3. Could antibiotics have been used in this baby to assist the therapy? Explain.

4. The physician in this encounter is a medical geneticist; yet this case appears to involve microorganisms. What is the lesson for physicians?

5. How can parents prevent this disease from occurring in their infants?

The epiglottis is the flap of cartilage that snaps shut in order to prevent food from entering the trachea and lungs when one eats. When the epiglottis becomes inflamed, it can block the airway and possibly cause suffocation, as this encounter explains. As we shall see, detecting an infection of the epiglottis is difficult and requires modern technology.

DISTANT ECHOES

TONY DAJER

It was a hopping, breakneck-paced Sunday evening in the emergency room. I had a patient's chart in my hands and a resident next to me, telling me her findings. "He has a sore throat," she said. I lifted my pen, ready to cosign the chart, authorize antibiotics, and move on to the next patient. But something in her voice stopped me.

"What is it?" I asked, immediately alert. Susan was one of the best residents in the hospital, with the sobering knack of usually being right when we disagreed on a diagnosis. If she was concerned, so was I.

"I'm not sure," she said slowly, frowning. "I think it's just a sore throat, but he has an odd, swollen lymph node under his chin." She shook her head and shrugged. "I don't know what to make of it. Could you take a look?"

I said I would, but before going over I asked a few quick questions: "Any fever? Pus on the tonsils?"

"Temp's 99.1. Tonsils are clean," she replied.

"Is the node soft like an abscess or hard like a tumor?"

"Neither. Just firm, and a little sore when you press on it."

The information was helpful, but I was really just buying time to think, the way all attending physicians do when smart young residents drop dilemmas on their doorstep.

Mr. Larma was sitting against one of the ER walls, his chair jutting halfway into a bustling hallway. Amid the controlled chaos of the emergency room, we didn't have the time or the space to give everyone a leisurely, secluded exam—especially if all you had wrong with you was a sore throat.

Even though he'd been waiting several hours, Mr. Larma smiled warmly as I walked over to shake his hand. He had a shiny bald pate, a stocky build, and the bluff, expansive manner of an Italian paterfamilias. If I hadn't seen his chart, I would have put his age at 60, not 71.

"Hello, Mr. Larma," I said, then nodded toward Susan. "I know you've already told Dr. Chen here some of what the problem is, but you know how we doctors are—we always need to hear it from the horse's mouth."

Mr. Larma opened his mouth to speak, swallowed with a sharp grimace, and then answered, "It's the throat, Doc." The words were muffled, as if his vocal cords had been wrapped in cotton. Gingerly, he raised a meaty hand to his Adam's apple. "She's pretty sore."

That tone of voice. Ten fire engines roaring through the ER, sirens clanging, couldn't have caught my attention as quickly. I had last heard that unique, muffled sound eight years before. But it reverberated as if I'd heard it yesterday.

Deliberately, I asked him the next question.

"Is this the worst sore throat you've ever had?"

"Absolutely, Doc. No doubt about it." Again that halting, almost disembodied sound.

His daughter, a fully coiffed but remarkably similar copy of her father, winked at him before piping up. "He loves to talk, doctor, but he's barely said a word all day."

"When did the pain start?"

"Early this morning," she answered for him.

It was now at least 12 hours later, and if my gut instinct was right, we might not have many more to fool around with. I turned to Susan and said, mildly, "Let's have a chat." Then I looked at father and daughter. "We'll be back in a minute, okay?"

Mr. Larma nodded appreciatively, as if we'd just taken his order for a gourmet dinner. His daughter sent us off with a cheery "You just take all the time you need, doctors."

Susan gave me an expectant look. I didn't mince words. "I think he has epiglottitis."

The epiglottis is perched mid throat, at the root of the tongue, where the trachea (which carries air to the lungs) branches off from the esophagus (which carries food to the stomach). It combines the beauty of an orchid with the prickliness of a Venus fly trap, and its job is to protect the trachea; it's a white petal of cartilage that lifts to allow air—and only air—to flow past. Routinely, it opens and closes thousands of times a day, snapping shut with each swallow to keep the delicate lungs free of any food, saliva, or microscopic aggressors you might swallow. It's the reason you can't drink and breathe at the same time. But sometimes the protector itself gets attacked by some nasty bacteria, and the resultant swelling—which we call epiglottitis—can completely close off the windpipe and kill a healthy adult in a matter of hours.

After hearing my diagnosis, Susan looked puzzled, then dubious. "You really think so?" she asked. "At his age?" The disease, as we both knew, most often strikes children between the ages of two and five, though it can be found in any age group. "And what about the lymph node? That's not a symptom of epiglottitis. There's no fever, and he has no trouble breathing. Shouldn't he, if he really has epiglottitis?"

Susan was my favorite devil's advocate, and I figured that if I could convince her, I could convince a consultant—something I'd eventually have to do if I was right, since the tests to confirm a diagnosis of epiglottitis can be performed only by a specialist.

"I agree that there's not much to go on," I began. "But he's having a lot of pain and it's not his tonsils—they're not even red—so it must be coming from farther down. And sure, the hoarseness could be due to simple laryngitis, but then it shouldn't hurt so much. I really think we've got to make sure it isn't epiglottitis."

Susan just nodded, slowly. No counterargument—a good sign.

"I've never seen a case."

"You just might have now."

In medical school, our professors used to thunder about the do-not-miss-this-on-peril-of-your-soul diseases. One of the trickiest, and most lethal, is epiglottitis. An adult's trachea at the level of the epiglottis is less than half an inch in diameter; it doesn't take much swelling to cut off the airflow in such a narrow pipe. In children it's even smaller, barely wider than one of their pinkie fingers. Mercifully, the disease is rare. But when it does occur, it can be deadly.

Epiglottitis is treacherous not because it lacks symptoms but because they're so commonplace. The hallmarks are sore throat and fever: What could be more innocuous? By the time stridor develops—the tight, organ-pipe whoop that's caused by trying to breathe through a severely narrowed trachea—it's almost too late. Unless a plastic tube is immediately inserted past the inflamed, swollen cartilage, the trachea will be closed off and air will no longer be able to get to the lungs.

So if Mr. Larma really was suffering from epiglottitis, we had to get moving. I told Susan to take him for a lateral neck X-ray—a view from the side that's the first step in diagnosing the disease. "Then I'll make some calls and see how hard it is to get hold of an ear-nose-and-throat guy on a Sunday evening," I added with a slight grimace.

While I had told Susan my provisional diagnosis, I didn't tell her why I felt so certain. I didn't tell her about the last time I'd heard that timbre of voice. Diagnosis by prior example is a lousy form of medical reasoning; it usually means a doctor doesn't know enough to consider similar, yet distinct, diagnostic possibilities. But with some prey, the scent never fades.

One memorable night during my internship, contrary to all the laws of medical probability, I'd seen not one but two cases of epiglottitis. It began innocently enough. A worried mother called me early on a Saturday afternoon. Her two-year-old was running a fever and complaining of a sore throat. I asked some routine questions. Was she drinking well? Yes. Any problems breathing? No, no problem. It didn't seem like an emergency to me, so I told the mother just to keep an eye on her daughter. Four hours later, I called back to check on her. The child just wasn't herself, said her mother, and the fever was climbing. I met them at the emergency room: a towheaded, strong-limbed child lay flat on her back, eyes half-closed, ominously indifferent to the hubbub swirling around her. She seemed to have been drained of her last drop of energy. A blood count and cultures, spinal tap, and chest X-ray gave me no clues.

Finally I examined her throat. I'll never know what made me save this routine maneuver for last, but when I did look I was startled. Popping up behind her tongue was a red, beefy epiglottis. It didn't make sense. Kids with epiglottitis, I'd been told, are usually sitting up and drooling, with their chins tilted forward to fight for air. They don't lie flat. But even though the case was unusual, it was clear that by putting a tongue depressor down the child's throat I'd broken a cardinal rule: never, ever, provoke an infected epiglottis. Like a Venus fly trap gone mad, the slightest pressure—even something as innocuous as the touch of a tongue depressor on the back of the throat—can send the epiglottis into an irreversible and instantly fatal spasm.

Feeling both lucky and sheepish, I approached the pediatrics resident. "I think I just saw a hot epiglottis," I said. He raised his eyebrows, skeptical, and went to see for himself. "Yup," he said, stick in hand, peering into her throat, "looks red to me too." Then he flinched. "I guess I shouldn't have done that," he mumbled. Gingerly, as if backing out of a minefield, we tiptoed from the room.

A breathing tube is routinely put down the throat of any child diagnosed with epiglottitis, since the airway is so narrow that any swelling at all will inevitably close it off. So we immediately called the anesthesia team. Within half an hour, the little girl had been put under, intubated, and pumped full of antibiotics. Relieved—and feeling very lucky that misleading symptoms and two impetuous young doctors hadn't caused any harm—I drove home. There was a message on my answering machine: "Call the adult ER. You have a patient with a bad sore throat."

"No way," I thought as I climbed back into my car. "Not two in one night."

The young woman sitting in the exam room didn't appear sick. I said hello. As she tried to answer, a beseeching look came over her face just before it contorted in pain. The "hello" she finally uttered was muffled and throaty. After that, each swallow or syllable brought a fresh spasm of agony, as if a branding iron were pressing on her throat. I carefully looked at her tonsils—no sign of infection. Her temperature was only 100.8, and her lateral neck X-ray showed no obvious "thumbprint": normally the epiglottis is a fine, feathery structure in an X-ray, but when swollen it looks thick and pudgy, like a thumb.

I was feeling cautious after my close call earlier in the day, though, so I called the ear-nose-and-throat consultant anyway.

Even after a long, painstaking exam with a laryngoscope (a narrow tube that lets a doctor peer beyond the tongue at the epiglottis and vocal cords), he was unsure whether the epiglottis was inflamed. Knowing neither of us would sleep that night if we sent her home, we admitted her to intensive care and gave her broad-spectrum antibiotics.

It was two days before we had proof that we'd done the right thing: that's when her blood cultures came back positive for *Hemophilus influenzae,* a bacterium that, despite its confusing name, is responsible not for cases of the flu (which is actually caused by a virus) but for most cases of childhood meningitis and epiglottitis. Happily, by then both she and the little girl (whose blood culture had also come back positive for the bug) were out of danger.

Eight years had passed since that Saturday. In that time I'd treated thousands of sore throats but not another case of epiglottitis—until now.

Twenty minutes after our somewhat one-sided chat with Mr. Larma, Susan had his lateral neck X-ray in hand. It looked suspicious but not definitive. I didn't care. My old enemy had never left very clear prints. Mr. Larma needed to be checked with an endoscope—an improved, longer version of a laryngoscope. In a hurry, I called Dr. Robertson, whose name headed the list of ENT specialists on call.

His answering service picked up the phone. "He's on vacation," a disembodied voice told me.

"Okay, then who's covering for him?"

"I'm not sure. Why don't you try Dr. Bernstein?"

I tried Dr. Bernstein. He asked for a quick description of the case. Despite my best effort to make Mr. Larma's plight sound dramatic, his only response was, "You're sure it's epiglottitis?"

"No, I'm not sure," I answered, beginning to get annoyed. "That's why I want him endoscoped."

"Listen, I'd be glad to come in, but I'm right in the middle of dinner, I'm not really on call, and I'm quite a way from the hospital. Why don't you try Dr. Wallace? He might be closer by."

I tried Dr. Wallace.

"Hmm, no fever and no respiratory difficulty?" he said. "You're sure it's epiglottitis?"

Once again I went through the whole song and dance.

"Tell you what," he said. "Why don't you see if the ENT resident is still in the hospital?" He gave me a pager number to try.

I glanced over at Mr. Larma, still sitting quietly in the hallway with his daughter, still looking as if he had all the time in the world. But I knew he didn't.

No one answered the pager number. Then Dr. Wallace called back.

"Any luck?"

"No," I answered curtly, letting a touch of exasperation seep into my voice.

"Okay, listen," he said. "I'll page the senior surgical resident and tell him where to find the endoscope so he can get started." (The endoscope is kept locked up in a special suite, available only to the ENT specialists, surgeons, and gastroenterologists trained to use it.) "I'll be in as soon as I can."

I let Susan know what had happened, and together we continued to wait. It took another half-hour. I attended to three or four more patients but kept a continual watch on Mr. Larma. He seemed to be holding his own. Finally the surgical resident arrived, dressed in scrubs and holding the long-awaited endoscope.

"He's over there," I said.

The three of us whisked Mr. Larma to our surgical room, where the surgeon removed the endoscope from its case. The quarter-inch-thick tube has an outer skin the color and texture of electrical tape. The resident fiddled with some knobs on the handle, and the tip of the tube bent preternaturally, like an alien proboscis straight out of the tavern scene in *Star Wars*. Mr. Larma seemed unperturbed.

"This will feel a bit uncomfortable, sir," the resident explained. "If it gets too bad, just say the word and I'll stop."

"You go right ahead, young man," Mr. Larma smiled, coolly eyeing the wriggling nozzle aimed at his nose.

After squirting novocaine jelly in the left nostril, the resident gently pushed the scope in, little by little, then stopped. He gave a low whistle and said, "There it is. No doubt about it."

He offered me the eyepiece. I got my anatomical bearings: the base of the tongue, the opening of the esophagus. Then I saw it—half-buried by engorged, cherry-red laryngeal tissue was Mr. Larma's epiglottis, twice its normal size. I passed the scope to Susan.

"I guess now I've really seen a case," she said softly.

Luckily, Mr. Larma's trachea wasn't dangerously blocked, so we didn't have to create an airway. Instead we admitted him directly to intensive care and blasted him with antibiotics, all the while keeping an intubation set open at his bedside, just in case.

Thirty-six hours later I paid a visit. Mr. Larma was sitting up in bed, looking genuinely tempted by his hospital breakfast.

"Hey, Doc, how ya doin'?" he boomed.

A moment of pleasant surprise: it was a completely different voice. Baritone. "I'm fine, thanks. So that's what you really sound like! You had us scratching our heads the other day."

"Well, Doc, looks like you hit the nail on the head anyway," he laughed.

The following day I went to drop off an instrument at the ENT office. While there, I ran into the chief of the department, who had taken over Mr. Larma's case. Since things had turned out so well, I decided not to mention how much easier it had been to convince Susan of the diagnosis than to persuade his colleagues to confirm it. Instead I simply asked him, "So what did you think of Mr. Larma?"

He became immediately stern, pushing a pair of bifocals up on his nose. "Obviously epiglottitis. The whole area was infected." His voice dropped to finger-wagging range. "You know, you people down in the ER have to be very careful. It's terribly easy to miss one of those. Nightmare if you do."

I practically laughed out loud, but stifled it. What could I say? Just that some voices are much more evocative than others.

UPDATE

Swelling of the epiglottis can be caused by burns from hot liquids, direct chemical exposure, and various viral and bacterial infections. As in this encounter, the most common cause of epiglottitis is an infection with Haemophilus influenzae *type b (Hib), the same bacterium that causes pneumonia and meningitis. Other bacteria, including* Streptococcus pneumoniae, *fungal yeasts, such as* Candida albicans, *and the virus responsible for chickenpox and shingles,* Varicella zoster, *can also cause an inflammation of the epiglottis. Routine Hib vaccinations for infants have made epiglottitis a rare condition, and it is now more common in adults.*

QUESTIONS TO CONSIDER

1. Why is this disease so dangerous in both children and adults?

2. In this incident, Mr. Larma did not display a high fever during the course of disease. What is the possible reason?

3. How might a physician distinguish a sore throat from an infection of the epiglottis?

4. Why is the disease in this encounter rare in humans?

5. What is the significance of the swollen lymph node observed under the chin of the patient?

*AIDS need not be a terminal disease as long as good med-
ical care and a firm resolve to remain healthy are parts of
the equation. In this encounter, two HIV-infected individuals attempt to fulfill their dream of having a child. In
doing so, they express their refusal to be defeated by HIV.*

BRAVE, BRAVER, BRAVEST

STEWART MASSAD

"I've been dreaming about having a baby," said Ashley.
That's not a strange thing for a childless 36-year-old woman
to tell her gynecologist, but it surprised me because this
patient is HIV-infected. Having a baby implies having a
future, something that those of us who have watched women
die of AIDS once never dared hope for.

But Ashley has always been tenacious. Six years ago, the
boyfriend who introduced her to heroin and HIV died of
pneumonia, leaving her resolved to convince others not to
duplicate her mistakes. She began speaking in schools,
women's shelters, and halfway houses, wherever she could
find an audience for warnings about unsafe sex and dirty
needles. Before long she met Ron, another former drug user
whose HIV diagnosis had shocked him into getting clean.
Their collaboration in the fight against AIDS inspired a love
bold enough to include the prospect of having a child.

When Ashley came under my care a few years ago, she
told me she had initially fought against AIDS with little hope
of victory. But in 1994, she participated in a landmark study.
The results demonstrated that a combination drug therapy
including inhibitors of the HIV protease enzyme, begun early
in the course of infection, could reduce the virus to unde-
tectable levels in blood samples and prolong life. Ashley

began to dream of having a child. She read all she could about her disease and the latest advances.

"I've been thinking about this for a long time," she finally said, as we sat together in a conference room, "but it seemed so unfair to bear a child I'd never be there for. Now, for the first time since my diagnosis, I feel like there might be a life for me. I want your perspective."

I took a long breath. Until recently, children born with HIV usually died as infants or toddlers, from diarrhea, pneumonia, or meningitis. Thanks to advances in antiviral therapy, those days are largely past, and children born with HIV now live into their teens and beyond. But all physicians who saw the 1980s, the first decade of the HIV epidemic, have memories they can't expunge—children isolated from other children, interacting only with parents who disappeared into addictions or died, doctors who had to abandon them for other patients and rotations, nurses who had to go on to other shifts and families of their own.

Those cases led some doctors to discourage HIV-infected women from bearing children. At that time, the odds of bearing an infected baby were unknown. To women who conceived unintentionally, the uncertainty was unbearable, and abortion often seemed the kindest choice.

In 1994, that uncertainty and fear began to recede. Results from the Pediatric AIDS Clinical Trials Group, a national consortium of clinicians and patients involved in experimental studies, showed that when mothers were treated with the anti-HIV drug zidovudine during late pregnancy and labor, only 8 percent of the babies were born HIV-infected. Among women given placebos, 26 percent of the babies were born HIV-infected. By lessening the amount of HIV in the mother's blood, zidovudine reduced the exposure of the baby to blood-borne virus during delivery. The results were so impressive that giving zidovudine is now standard for all pregnant women infected with HIV. In addition, obstetricians now encourage HIV testing for expectant mothers to prevent unwitting maternal transmission of the virus.

More recent studies have deepened our understanding of anti-HIV drug therapy. When mothers take zidovudine along with protease inhibitors and other drugs, the risk is much lower than when they take zidovudine alone. In Ashley's case the infection was relatively well controlled. At 397, her CD4 count, the number of infection-fighting T cells, was OK, and the virus in her blood was too low to show up in tests. Staying healthy required a complicated drug regimen: pills taken up to five times each day, some with food, some on an empty stomach. But Ashley was nothing if not dedicated.

"Times have changed," I told Ashley. "The odds have improved, but your child could still be born with HIV and never live a normal life. And to say that the odds of an infected baby are 1 percent or 3 percent doesn't mean the baby would have a 1- or 3-percent infection: It's all or nothing. Can you live with that?"

She nodded. "It's like playing Russian roulette with a gun that has 30 chambers: If I'm unlucky, I'll still get blown away." She stood up. "I'll let you know."

I didn't see her until she returned six months later for her annual Pap smear. "We're trying," she told me. "Some people tell me that it's selfish, that any risk of handing on this disease is too great. But to me, having a baby is standing up to the virus. We think it's time to look ahead, to create life, not just avoid dying."

We talked about the ethical issues her pregnancy raised. We talked about how HIV mutates rapidly, and the same viral strain can evolve differently in different individuals with different immune systems and genetic makeups. Ashley accepted that if Ron stopped using condoms it was theoretically possible that she might get infected with a more virulent strain of HIV. She understood that in the months or years to come the virus she carried might become resistant to drugs, killing her before her baby had a chance to know her. She had made plans: Her sister had agreed to raise the child if necessary.

Within three months, Ashley was pregnant. She faced morning sickness, which compounded the nausea caused by her anti-HIV drugs. Twice she had to be admitted to the hospital because she could not keep any liquids or anti-retroviral medications down. But with an antiemetic patch behind her ear, she managed to keep taking her pills, even when she could swallow nothing else.

By the fifteenth week of pregnancy, Ashley's appetite returned. The next issue she faced was amniocentesis. For pregnant women of her age, genetic testing is standard to identify babies with Down syndrome and other defects. But the needle required to draw cells from the amniotic fluid can introduce HIV into the fetus. Ashley decided to forgo the procedure, as HIV-infected mothers are advised to do, and get a detailed ultrasound instead. Her knuckles were white as she clutched Ron's hand during the procedure, but the ultrasound was fine.

Ashley's contractions started early during her third trimester. Soon after, she quit work to rest in bed. Weekly ultrasound scans showed the baby growing and kicking, stretching in anticipation of birth.

How to deliver the baby was the next critical issue. Cesarean section reduces the risk of HIV transmission to the child because the baby encounters the mother's virus-infected blood only briefly during the procedure. Unfortunately, the risks to the mother—infection, bleeding, and anesthetic complications—are higher for cesareans than for vaginal deliveries.

"All that matters is protecting my child. I'll take the knife," she said.

Eight months into the pregnancy, Ashley's contractions picked up again. Her cervix softened and began to open. We had to perform the cesarean before the bag of amniotic fluid broke, exposing the baby to virus in the mother's body.

Like any surgical procedure, cesarean section requires all the usual precautions: gowns, masks, gloves. But in the age of AIDS, we have added new barriers. The masks have shields to protect against splashes of blood. The gowns are impermeable. Everyone on the operating team wears two sets of gloves, and shoes are covered with knee-high gaiters. Still, performing a cesarean section on a woman with HIV is frightening because the initial focus is not to stop the bleeding but to deliver the baby as quickly and safely as possible. Removing the placenta is especially worrisome, because potentially lethal blood mixes with amniotic fluid and spills over the operating drapes.

Ashley's surgery was uneventful, and her baby girl, although small, seemed to thrive. She longed to nurse her daughter, but she had to give her formula to avoid transmitting the virus through breast milk. When the time came to test the baby for HIV, Ashley burst into tears. When the test results came back, she cried again. The child had escaped infection.

The parents named her Hope.

UPDATE

Since 1985, the number of AIDS cases diagnosed among women has more than tripled, from 8 percent in 1985 to 27 percent in 2005. However, during these past 20 years many women living with HIV have come to enjoy much longer and stronger lives. In fact, with proper care and treatment, many HIV-positive women can continue to care for themselves—as well as their babies. The One Test. Two Lives. *program from the Centers for Disease Control and Prevention (CDC) focuses on providing screening materials to doctors to ensure that all women are tested for HIV early in their pregnancy because perinatal transmission accounts more than 90 percent of all AIDS cases among children in the United States. As illustrated in this encounter, antiretroviral therapy during pregnancy can substantially reduce the transmission rate of HIV. In 2007, the CDC estimated that out of 50 pregnant women with HIV, the risk of passing HIV to their babies is about: 1/50 (2 percent) if women begin antiretroviral treatment during pregnancy; 5/50 (10 percent) if women begin treatment during labor, or their babies get treatment soon after birth, or they both get treatment at these times; and 13/50 (26 percent) if women receive no treatment.*

QUESTIONS TO CONSIDER

1. What were the probable sources of HIV infection in Ron?

2. How did Ashley's determination to beat HIV help keep her baby free of the virus?

3. What special precautions were taken during the cesarean section procedure used to deliver the child?

4. Ashley wished to nurse her baby but was advised against doing so. Why?

5. After having their child, Ron and Ashley should return to using condoms even though both are already infected with HIV. Why is this necessary?

Medical professionals don't always agree on the best ther-
apy for treating an infected patient, and it may be neces-
sary for a physician to take the unusual step of bypassing
the normal channels of authority, especially if the life of the
patient is believed to be in danger. In this encounter, a
physician helps the patient even though he risks criticism
from his colleagues.

A LETHAL SCRATCH

TONY DAJER

"This leg is hot, doc," the ambulance medic whistled.
"Scratched the bottom of her foot three days ago. A nick.
Nothing." He threw up a perplexed hand.

Mrs. Anders, a 35-year-old black woman, brooded in her
wheelchair. Vexed at the absurd injury that had landed her in
the emergency room, she just wanted her leg to get better so
she could go home to her three kids.

"Go to it," I told Kevin, my physician's assistant student.
Kevin was a star, the kind of PA student who sometimes
makes medical school seem overrated.

"Get the history, examine her, then tell me your plan," I
told him, unnecessarily.

Ten minutes later he crisply read out of his little spiral note-
book, "The patient scratched the bottom of her left foot three
days ago. Can't remember what on. Yesterday it began to hurt
and swell. Last night she felt feverish. Except for a history of
depression, she has no other medical problems. No diabetes,
no prior leg surgeries, no nothing, really. Her temp is 104.6."

"Whew," I exhaled.

"On exam," he continued, "the left foot and ankle are red,
hot, and tender."

"Any other findings?"

"None."

"What do you think?"

"Cellulitis of the left ankle and foot. Needs antibiotics."

A superficial skin infection, cellulitis is both common and generally easy to treat. The usual causes, staph and strep bacteria, respond to oral antibiotics. Doctors prefer to treat both because it's difficult, clinically, to distinguish them.

"Sounds good. Let's take a look."

The foot and ankle glowed a dusky red. But Mrs. Anders still had her pants on.

"How far up did you look?" I asked Kevin.

"Oops," he replied. "Not far enough."

I smiled at Mrs. Anders and drew the curtain.

"We'll need to take these off, I'm afraid."

"Okay," she answered neutrally.

We helped her slide the slacks off. And there, tracking up the inside of her leg like a battalion of fire ants, was an inch-wide streak of red.

I pointed: "Lymphangitis. Classic sign of strep infection. The bug uses the lymphatic system like a highway. This infection has advanced well beyond her ankle. She needs intravenous antibiotics."

"And I should have picked it up," Kevin lamented.

"No, I think an infected leg is enough for now," I reassured him.

The lymph system is the complex network of vessels that carries waste products from the tissues back into the bloodstream. These vessels also carry foreign material to the lymph nodes to help promote an immune response against infection. The worry with strep infections in the lymph system is that they can linger in the vessels, causing local destruction before the immune system can be alerted to launch an effective defense. Alternatively, if strep passes out of the lymph system and invades the bloodstream, it can cause catastrophic illness without much local infection. In Mrs. Anders's case, strep was on the march and we would have to hurry to head it off.

Streptococcus bacteria, the cause of everyday strep throat, is among our most ancient, tenacious, and versatile bacterial enemies. Strep strains linger in the soil and on our skin, and their tricks are legion. Some can incite surface infections like strep throat and erysipelas. (The word *erysipelas*—Greek for

"red skin"—refers to a more severe skin infection than cellulitis.) In other cases, they can invade lungs, heart valves, and
spinal cords. A great opportunist, strep rampaged through
nineteenth-century maternity wards as puerperal fever thanks
to doctors who examined new mothers with unwashed hands
between cases. Before penicillin, untreatable strep throat
caused lethal epidemics of rheumatic fever via an evil bio-
chemical mimicry. The strep provokes an antibody response
that mistakes the sufferer's own heart muscle and valves for
the strep intruder.

Most recently, strep reared its hydra-like head as "flesh-
eating bacteria." Headlines hyped it as a "new" plague, but
2,500 years ago Hippocrates described an erysipelas that led,
gruesomely, to "flesh, sinews, and bones falling away in large
quantities." This horrifying condition is caused by strep
strains that slip into tissue following an innocuous scrape or
cut. If the infection is not treated, the bacteria can crank out
enough enzymes and toxins to literally dissolve flesh.
("Tissue necrosis" is the official term.) Strep enzymes can dis-
mantle connective tissue, blood clots, and other living fire-
walls in its path. Strep toxins sabotage blood vessels and cell
membranes, dropping blood pressure and flooding organs
with oxygen-blocking cellular sludge. So efficient and tailor-
made to humans are strep's tools that in one of biology's
great ironies, we now use the enzyme streptokinase as a clot-
buster to open clogged coronaries and save tens of thousands
of heart attack victims a year.

Finally, this bacterial version of a Swiss Army knife kills
because it is quick. Once it clears a beachhead, it can move at
almost visible speed, as it had on Mrs. Anders's leg. The nas-
tier strains of strep secrete flesh-chewing toxins, and if the
infection progresses, patients can suffer the horrendous loss
that Hippocrates so accurately described. The destruction of
flesh is even more terrifying because it can commence without much evidence of poisoning on the surface of the skin.

"So what do we give her?" I asked my still-chagrined PA
student.

"Cellulitis can be caused by staph, strep ..."

I cut him off. "Usually. And you need antibiotics that cover
both. But," I deepened my voice and harrumphed, "this is classic, absolutely classic strep lymphangitis. Only strep behaves

in such a fashion. Your patient requires penicillin. And in large doses. So fire away."

In keeping with its particularities, strep, in this era of antibiotic resistance, has remained exquisitely sensitive to that granddaddy of antibiotics, penicillin.

"A million units?" Kevin asked.

"Her temp is over 104, the strep is galloping up her leg," I replied. "Let's say 2.5 million."

All that remained was to call the admitting team and get Mrs. Anders upstairs. I let Kevin do it.

"And emphasize to them it's strep and she needs penicillin."

"Right."

We moved on to other patients. An hour later, I buttonholed Mrs. Anders's admitting intern, Carol Fields.

"Amazing lymphangitis, eh?" I asked. "You won't see many as clear-cut as that. How much penicillin did you give?"

She squirmed.

"Our attending wants Cefazolin," she finally admitted. "Broader coverage."

Cefazolin is distantly related to penicillin; both muck up the bacterial cell-wall-building machinery critical to cell division.

What Cefazolin has over penicillin is its resistance to staph's penicillin-chewing enzyme, penicillinase. But against strep, penicillin leaves Cefazolin in the dust, clobbering the bug at concentrations of only ten-millionths of a gram per liter. As horses, Cefazolin would play Clydesdale to penicillin's Thoroughbred. And in Mrs. Anders's case, we had a lot of lost ground to win back.

"But it's classic strep," I said, flipping my hands up in supplication. I pulled out *Rosen's*, the emergency medicine bible.

"See, right here, 'streptococcal ascending cellulitis.' This is a carbon copy of your patient."

"I know, Dr. Dajer, but I'm only the intern."

I headed off and snagged the resident.

"Look," I began, "normally I'd agree with you on the Cefazolin for added staph coverage, but this is strep. The trivial entry wound, the lymphangitis. It all adds up. She needs pen."

"We'll stick with the Cefazolin for now, thanks," he replied in a tone that dripped "This is routine, it's only strep or staph."

Every week, medical journals plug newer and broader-spectrum antibiotics. The temptation to believe in the latest sawed-off shotgun is fueled by tens of millions of dollars in annual drug company profits. And strep, still quaintly sensitive to penicillin in an era of mega-resistant, armor-plated bacteria and implacable viruses, seems a pushover, a has-been you can practically pick off with a slingshot.

"Broader isn't deeper. Pen is the drug of choice," I persisted.

The resident smiled stiffly. He was busy. "It's what our attending wants."

I gave up. I couldn't prove it was only strep, because lab cultures take two days. Moreover, medical etiquette demanded I pass the baton to the admitting team.

Next morning I sent Kevin up to check on her.

"She looks better," he reassured me. "Fever's down to 102, and she says it hurts less."

"Ah well," I sighed, "so much for the classics. She looked sick when she came in. I'm glad they were right."

On the second morning, Kevin needed no prompting to run upstairs. When he came back down, the look on his face chilled me to the heart.

"I don't think she looks good," he said with an urgency I'd never heard him use before. "The medicine team says in its note she's better, but her temp's back up to 103.6 and the leg looks terrible."

I sprinted up the stairs. At each landing my brain pounded with the thought, "Christ, two days. Two days wasted." Mrs. Anders's only reply to my out-of-breath "How are you?" was a moan. The leg had swollen enormously and taken on an ominous purple hue. Small, evil-looking blisters puffed across the skin. I feared tissue necrosis, the beginning of catastrophe.

Once strep succeeds in penetrating deeper tissue layers with its cutting and liquefying enzymes, a chain reaction begins: muscle cells killed by strep release potassium, phosphate, and other cellular by-products that poison adjacent muscle. Strep feasts on the remains and oozes new waves of deadly toxins. At that point the only hope for survival is to flay open the limb and excise wide swaths of flesh.

But even with drastic surgery, strep can outstrip its pursuers. Patients can lose not only swaths of skin but limbs too.

Carol, the intern, sat at the nurses' station jotting down lab results.

"Your patient is worse," I announced bluntly.

"But the resident said her white count was down and she looked better," Carol replied.

Then she held up a lab slip and said, "Why do you suppose her PTT is up?"

The PTT, or partial thromboplastin time, is a measure of how well the clotting system is working: the higher the number, the poorer the function. The test involves taking a bit of blood, adding in a clotting factor, and measuring how long it takes for a clot to form. Mrs. Anders's result hinted at a battle in her blood. Strep, in addition to killing tissue directly, can secrete toxins that can wipe out kidney, lung, and coagulation function, for starters. Now my alarm bells whooped.

"Because she's in the early stages of toxic shock or tissue necrosis or both," I wanted to shout. Instead I said, "You are to give her 3 million units of penicillin right now and call an infectious-disease consult. Stat."

"But I can't do that," Carol replied. "It's up to my attending."

"Do it and I'll deal with your attending," I said with heat. The trick to medical argument, paradoxically, is to sound absolutely sure of yourself, though it's a state achieved only by the deeply ignorant. I called the attending.

"Dr. Moore, this is Dr. Dajer. I admitted Mrs. Anders from the ER. Her leg looks much worse to me. I'd like to have the intern start high-dose penicillin and call an I.D. consult. Her PTT is up. She may be in the early stages of toxic shock or tissue necrosis."

"But she looked better this morning," Dr. Moore stammered.

"But much worse than she did two days ago," I insisted.

"Shall we see what the I.D. consult says?" she temporized.

"Yes, but we should start pen now."

"I'd rather wait." Medical convention says that cellulitis is easily treated with Cefazolin, and Dr. Moore was sticking with convention.

"Fine," I said, then rang off. I turned to Carol.

"Give her the penicillin. Don't worry, I.D. will agree."

The infectious-disease specialist was somewhat startled by the "stat" consult in a field that usually counts hours or days, not minutes, in the onset of disease. But she took one look at Mrs. Anders's leg, upped my penicillin dose, and added Clindamycin. For strep has another trick: once it has reproduced in large numbers, it more or less stops multiplying. At this stage cell-wall monkey wrenches like penicillin aren't as effective because few new bacteria and new cell walls are being made. Clindamycin tackles this phase by directly inhibiting the bacteria's protein-making life-support systems. Moreover, toxins and destructive enzymes are proteins, so Clindamycin offers the theoretical advantage of shutting down the poison factory as well. But I.D. was so worried she called the surgeons. The surgeons ordered a CT scan.

If it showed tissue destruction, I would never forgive myself. Any mottled or ratty-looking patches would be a sign that strep toxins were tearing up the normally smooth and dense muscle tissue.

But the CT scan was negative. The next day Kevin and I trooped up together. The leg still looked bad, but no worse. Mrs. Anders lay watching TV, ignoring us, as if to deny any association with that damn leg.

The following morning, Kevin came downstairs looking worried.

"It's bad," he intoned. "The leg has huge blisters filled with pus. It's like all her skin is falling off."

But I had been upstairs before him.

"No," I reassured him, "it's not pus, just yellow inflammatory fluid, as in a blister after a burn. She's getting better. But if you hadn't figured out what was happening, she'd be dead now."

Kevin didn't smile, despite the compliment. He still seemed stunned at how unforgiving a "routine" infection could be.

A week later I encountered the I.D. consultant.

"You know," she told me, "the medicine team still says the PTT was a lab error. They think she was doing fine on the Cefazolin. They said it was only strep."

Strep has a long and daunting history of capitalizing on human inattentiveness. Where lies the pigheadedness gene, I wondered, that deludes each generation into dismissing the lessons nature has taught its forebears—and at such cost?

The consultant shrugged. "Kids these days. No respect."

UPDATE

Cellulitis is an inflammation of the skin layers below the surface; that is the dermis and subcutaneous tissues of the skin. The infection often starts after bacteria, most often Streptococcus *or* Staphylococcus, *enter through a crack or break in the skin. Cellulitis is most commonly found in the legs, especially near the shins and ankles. As is evident from this encounter, if the bacterial infection is left untreated or misdiagnosed, it may rapidly turn into a life-threatening condition. For this reason, early diagnosis and treatment of cellulitis is important.*

QUESTIONS TO CONSIDER

1. Under what circumstances might Mrs. Anders have become infected?

2. This encounter presents many possible avenues by which the infectious agent can damage body tissues. Describe some of the methods.

3. In this encounter, why is a broad-spectrum antibiotic described as a "sawed-off shotgun"?

4. One effect of infection by the streptococci is "toxic shock." Which other well-known bacterial species can also cause toxic shock?

5. Why is penicillin less effective on bacterial cells that have slowed down their division rate?

In numerous incidents each year, a patient is admitted to the hospital with a fever of unknown origin. Then the medical detectives are called into action to develop a diagnosis and to prescribe the correct treatment. In this encounter, we see how a hospital pathologist takes the available evidence and narrows the possibilities before finally identifying the disease in his patient.

MYSTERIOUS FEVERS

DANIEL C. WEAVER

It was full summer in southern Indiana. The morning mist rose through the oaks like steam. Bob Ervin was so hard at work cobbling together his daughter's playhouse in the backyard that he didn't notice anything unusual.

He finished the playhouse and felt fine—for a week or so. Nine days later he woke in a sweat. Unable to return to sleep, he got up at 3 A.M. and turned on the light. The pain in his eyes was so sharp it felt like a stabbing knife. He flipped the light off and spent the rest of the night in a feverish sweat.

The next day he went to the family doctor. An examination revealed a fever of 102. Except for some blood in his urine, nothing was unusual. The doctor gave him Bactrim, an antibiotic commonly used to treat urinary infections.

During the next week, Mr. Ervin dutifully took the antibiotic, but the fever never broke. By the end of the week he was tired and weak; he came to the emergency room with a temperature of 103.5. The emergency room physician's examination failed to reveal any sign of infection, and the chest X-ray was clear. Routine blood cultures and chemistries were drawn and sent to me in the hospital's pathology lab.

At that point, Mr. Ervin was sweating so much he was at risk of dehydration. The emergency room physician decided to admit him with the diagnosis "fever of unknown origin."

Every medical student learns about this type of fever. There are several well-established strategies for figuring out the cause, but none are definitive. They can't be. There are literally thousands of causes, from a swallowed toothpick stuck in the gut to cancer to a blood clot to some strange contagion from an exotic land. The clue often lies in something overlooked in the patient's history.

Just after Mr. Ervin was admitted, his family doctor visited and combed the history for a hidden cause one more time. Mr. Ervin denied a host of symptoms: coughing, vomiting, shortness of breath, pain on urination, recent weight loss. The only new details were vague complaints about discomfort after eating. His doctor then reviewed the initial laboratory results. The routine blood cultures for infections were all negative, but the blood count didn't look normal. Blood counts tell whether three types of cells—white, red, and platelets—are in correct proportion. Mr. Ervin had low levels of all three, a condition called pancytopenia. And the levels of his liver enzymes—which increase in the case of infection or early liver damage—were high. He had anemia, weakness, and fever.

Trying to unite as many of the symptoms as possible in a single diagnosis, the doctor concluded that Mr. Ervin's symptoms spelled acute cholecystitis: an inflamed gallbladder.

The gallbladder, roughly the size of a kiwifruit, lies just beneath the liver. Its job is to store bile produced by the liver and secrete it whenever fat needs to be digested. When the gallbladder is acutely inflamed, it's usually because stones—clumps of fatty material or bile pigment—have accumulated inside it. A patient who has a gallbladder problem will often have fever, elevated liver enzymes, and some discomfort after eating.

An ultrasound of the abdomen revealed signs of gallstones. Audrey Nevins, a surgeon, was called in for a consultation. She couldn't feel the stones, but that didn't mean Mr. Ervin didn't have them. Gallstones can be difficult to detect because of their location behind the liver. Still, Nevins felt that something didn't fit the diagnosis: the pancytopenia.

Problems in the gallbladder aren't associated with drops in blood count.

"Have you been visiting anywhere?" she asked.

"Yeah, Colorado. I was hiking last month."

"Were you bitten by any animals or bugs?"

"A few ticks here and there."

"What do you mean 'here and there'?"

"Some in Indiana, some in Colorado."

The surgeon left the room and came to see me. In a large medical center, a case of fever of unknown origin would be handled by an infectious disease doctor, a superspecialist who knows the sexual habits of the tsetse fly and is intimate with disorders such as loa loa and looping disease. But here in rural Indiana, the doctors call me. I'm the pathologist, and I have the books.

Nevins said, "Mr. Ervin went hiking in Colorado last month, and he had tick bites."

"Sounds like a tick-borne disease," I said, and began paging through my family medicine textbook. Ticks transmit hundreds of diseases around the world, and in the United States they are the most common vector of infectious diseases. Several tick-borne diseases—Rocky Mountain spotted fever, babesiosis, tularemia, and Lyme—are found here, so it would be easy to narrow down the exact cause. Ticks harboring a bacterium called *Rickettsia rickettsii* transmit one of the most common tick-borne diseases, Rocky Mountain spotted fever. Mr. Ervin's signs and symptoms seemed to fit that diagnosis: the pancytopenia, the fever, the elevated liver enzymes, and the tick bites.

"Does he have any spots?" I asked.

"Not that I could see," said Nevins.

I quickly scanned the description in the text and found that a small percentage of patients with Rocky Mountain spotted fever do not have spots.

"How do we confirm it?" Nevins asked.

I grabbed the current pathology textbook to see what I could find on diagnostic tests for the disease. She glanced into the text I was reading.

"What about that one, the one in the next paragraph— *chaffeensis*."

I skimmed the passage. Lone star ticks infected with the bacterium *Ehrlichia chaffeensis*, transmit ehrlichiosis, which was first diagnosed in humans in 1986. Its full name is human monocytic ehrlichiosis because the bacterium thrives inside white blood cells called monocytes. Since *E. chaffeensis* doesn't grow in any standard laboratory media, blood cultures for the infection turn up negative. The disease usually shows up in middle-aged men with recent tick bites, and their symptoms are typically fever, pancytopenia, elevated liver enzymes, and photophobia, or extreme light sensitivity. I had heard of ehrlichiosis, but I wasn't aware of cases in southern Indiana.

"Audrey, it looks like a perfect fit," I said. "Tell him that tetracycline will probably clear it up easily."

She headed out to talk with Mr. Ervin. Now it was my job to confirm the diagnosis. First I called a medical microbiologist at Indiana University. He'd never seen an ehrlichiosis case and suggested I call the Centers for Disease Control and Prevention in Atlanta. A few phone calls later, I learned of an antibody test that could identify it. The problem, though, was that it couldn't tell if the infection was current. I needed something better.

I logged onto the National Library of Medicine's Web site. In a rural hospital the Internet can be a lifeline to facts. A list of 20 articles on ehrlichiosis popped up. One described a new method of testing for the pathogen's nucleic acid in the patient's blood. The author was Stephen Dumler, a medical microbiologist at Johns Hopkins. After a few phone calls, we found out what kind of blood sample to prepare, drew Mr. Ervin's blood, and sent it off to Dumler. Two days later, he called to confirm that the test had detected the presence of *Ehrlichia chaffeensis.* We had nailed it.

Meanwhile, the surgeon had started Mr. Ervin on a week-long course of antibiotics. His temperature returned to normal.

But there's an epilogue to Mr. Ervin's story. The month after we diagnosed his condition, researchers at Ball State University in Muncie, Indiana, published a study showing a high rate of ehrlichiosis among lone star ticks in southern Indiana. Some researchers believe the incidence is on the rise because of changes in the region's ecology. Oak and hickory forests are growing back, and that may be boosting the number of deer, the tick's major host. According to the Ball State study, my hometown, Jasper, Indiana, is the apparent epicenter of this infestation.

With skills honed over 300 million years of evolution, the tick is a formidable foe. In the warm seasons it lies waiting for a meal of fresh blood. It alights, walks about, selects its site, unsheathes its knives (called cheliceral digits), and slashes into the victim's skin. Then it inserts its hypostome, a sword-like probe with row upon row of microscopic teeth, and begins to suck. Hours to days later, fully engorged and sometimes 100 times heavier, the tick falls off. In the meantime, there has been a small transfer of fluid from the tick to the victim. If the tick is infected with a pathogen like *E. chaffeensis*, it can transmit the infection in that fluid.

Sometimes ehrlichiosis is so mild that cases can pass like the common flu, so it's hard to know how many cases are missed. Some researchers suspect most are. If the disease is recognized and treated with tetracycline, it's easily cured. But if it's missed and infection progresses without treatment, patients can die from causes ranging from brain infection to respiratory problems.

Nearly a year has passed since Mr. Ervin encountered that tick in the woods of southern Indiana. It is now, once again, full summer here, but this time we know the beast.

UPDATE

In the United States, human diseases caused by Ehrlichia *species have been recognized since the mid-1980s. As described in this encounter, human monocytic ehrlichiosis (HME) is caused by* Ehrlichia chaffeensis *and is transmitted by the lone star tick— and possibly other ticks. The white-tailed deer is the major host and natural reservoir for the bacterium. The Centers for Disease Control and Prevention (CDC) have reported more than 400 HME cases per year in the United States with most occurring in Missouri, Oklahoma, Tennessee, Arkansas, and Maryland. Often clusters of HME are reported. In 2007, there were more than 150 cases reported just in Missouri. Sadly, a 15-year-old Missouri girl died from HME in August 2007, just two days after being admitted to the local university hospital. To help prevent tick bites, people should avoid wooded and bushy areas with tall grass or wear long pants and long sleeves, and use insect repellants that contain 20 to 50 percent DEET.*

QUESTIONS TO CONSIDER

1. Considering the mode of transmission in this encounter, what other diseases could have caused Mr. Ervin's illness?

2. What is the prognosis for an individual not treated for the disease discussed in this encounter?

3. How does Mr. Ervin's case show the relationship between the infectious disease and the ecology of a region?

4. Why couldn't the infectious agent be grown in a blood culture?

5. From the information provided, can you guess how Mr. Ervin first became infected with the disease?

Sometimes a diagnosis leads a physician through a maze of possibilities and dead ends before the physician believes the mystery has finally been solved, Then the lab results show up, and they reveal that the patient is suffering a completely unexpected and different illness than originally believed. This encounter recounts such a situation.

TRIUMPH BY TREACHERY

TONY DAJER

"*Ay, me duele, doctor. Por favor, ya no.*" It hurts, doctor. Please, no more.

My right hand recoiled instinctively, but I had to be sure. It sank carefully—again—into the soft right lower part of my patient's belly. Again the 57-year-old Puerto Rican woman cried out.

No doubt about it. This was classic appendicitis. I put my left hand—the innocent one—on her shoulder.

"*Señora,* it's very possible," I began, in Spanish, "that you have appendicitis. You may need an operation. In a little while the surgeons will come down to see if I am right. Until then—and you must forgive me—we can't give you anything for the pain. If we did, it would make it very difficult to arrive at the correct diagnosis."

"*Ay, doctor,*" she moaned, not very comforted by my clinical imperatives. "*Por favor, haga algo.*" Do something, doctor.

I squeezed her shoulder and mumbled, "As soon as I possibly can."

At the nurses' station I picked up her chart to make sure I had all my ducks in a row. Her temperature was 101.8. Her white blood cell count—a reliable sign of infection—was 19,000, far above the normal range of 4,300 to 10,800. Her abdomen was tender right over McBurney's point, the spot about midway between the navel and the comer of the pelvis that people commonly call the hipbone. But the clincher was that Mrs. Velez complained of pain even when I tapped gently over her McBurney's point. This response, called rebound tenderness, is triggered when the outer covering of an inflamed appendix rubs against the nerve-rich wall of the belly. My patient needed to be in the operating room.

The surgeons answered my page right away.

"I think I've got a hot one for you," I said, rattling off Mrs. Velez's symptoms of fever, high white blood cell count, and tender belly.

"Yep, sounds like the real McCoy," the chief resident drawled. "We'll be right down."

A throng of blue-pajamaed residents materialized in the emergency room.

"Over there," I said, nodding toward Mrs. Velez's cubicle.

I watched them troop in, expecting an instant verdict and a swift decampment to the operating room. Instead they just stood inside, conferring, reexamining, milling about like a football huddle during a time-out. Finally the chief emerged.

"You know, I agree she's tender, but she doesn't need an operation," he said, almost apologetically. "I'd guess her exam has changed since you saw her. To me, she's more tender on the left—and higher up. And she says she's been having a lot of diarrhea. I think she's got enteritis."

Enteritis means inflammation of the intestine, but it's far from an exact diagnosis. It's a bit like saying "headache" to explain pain above the neck.

"Enteritis?" I squinted dubiously.

"Yeah. I can't give you a better answer than that. She definitely needs to be admitted—the high white count worries me—but I also know she doesn't need an operation."

Sheepishly, I went back to Mrs. Velez. Sure enough, her belly was now tender on the left. From certainty to bafflement in 30 minutes. "Seven years of training just to be confused," I muttered to myself. "And eight years' experience," piped another little voice in my head. Nothing to do but start asking more questions.

"Mrs. Velez, it appears you may not have appendicitis," I began. "That's the good thing. The bad thing is that now we're not sure what you have. Do you think you could answer a few more questions?"

Mrs. Velez pressed gingerly on her abdomen. "Well, it seems less painful. So maybe I'll remember better this time."

"Very good," I said. "Let's see, the only medicines you take are for asthma, right?"

"Yes. The Proventil inhaler and, when I need them, steroid pills."

Inhalers, the standard treatment for asthma, relieve wheezing by delivering an agent that dilates the bronchioles, the tiny airways that constrict during an attack. The pills, which contain steroid hormones, suppress the migration of white blood cells and hinder the release of the inflammation-causing substances that clog the airways and ignite asthma flare-ups. Because long-term steroid use damps the immune response, doctors reserve steroid pills for severe asthma attacks. Mrs. Velez had averaged three or four courses of steroids a year. She had finished the last one three weeks before.

"And that's it?" I asked.

Mrs. Velez folded her hands over her stomach. "Well, two weeks ago I finished a series of antibiotics. The doctor said I had bronchitis."

"Oh really? Which antibiotic?"

"Big white pills."

"How many times a day?"

"Twice."

"Do you remember if they were called Bactrim, by any chance?"

Her hands flew up in confirmation. "Yes. That's what they were called."

"And your diarrhea started when?"

"Yesterday. But ay, doctor, today it's very bad."

"Watery or bloody?"

"Watery. Lots and lots of water."

A light went on in my head. But I still had a few more questions.

"And you've never been sick from anything else?"

"No."

"No recent travel?"

"No."

"And you're from Puerto Rico, right?" I continued.

"Pues claro, of course," she grinned. "From Utuado. The mountains. And from your accent, *doctorcito,* it sounds like you're from the coast."

"I confess," I smiled back. "I grew up in San Juan."

"But you know, I'm really from here. From New York. I came when I was 15," she corrected, then gave me a nudge. "I've probably been living here since before you were born, right?"

"Right again, Mrs. Velez."

"*Pues, mijo,* what are you going to do about my problem belly?"

"I think I may have just the trick," I winked.

Feeling like a doctor again, I sauntered over to the surgery resident.

"Did she tell you about that Bactrim two weeks ago?" I asked him.

He looked up from the note he was writing. "You know, I was thinking the same thing."

"You were? So you've thought of pseudomembranous?"

"Yup," he nodded. "Definitely a possibility."

Pseudomembranous colitis is one of the terrible reasons that antibiotics should not be prescribed without good cause. Common antibiotics like ampicillin and Keflex (and, less commonly, Bactrim) can wipe out the colon's normal bacteria and allow a nasty bug called *Clostridium difficile* to run rampant—even up to six weeks after the antibiotics are finished. In severe cases, the two toxins secreted by *C. difficile* can cause the lining of the gut to slough, creating a "pseudomembrane" that can be seen when the colon is examined with a fiberoptic scope. Ironically, ridding the colon of *C. difficile* requires another antibiotic—Flagyl or vancomycin.

I told the admitting team about my now-not-so-original idea and ordered a full series of stool tests for Mrs. Velez. I then stopped in to say good night to her. I left expecting to find a much happier patient the next morning.

It was not to be. The first person I met in the hospital the next day was Mrs. Velez's son.

"Doctor," he pleaded, "she hasn't slept all night. She's been up every ten minutes with the diarrhea. It never stops."

My heart sank. Mrs. Velez should have improved a little by now. And to make matters worse, she was still in the ER holding area, waiting for a more comfortable bed upstairs.

"*Ay, doctorcito,*" she complained, "I'm like an open faucet. You must do something."

I reviewed her orders. Flagyl, the antibiotic that kills *C. difficile,* had been given by vein instead of by mouth. This was not a mistake, but, I reasoned with the residents now caring for Mrs. Velez, better to put it directly into the intestine, where it was needed.

"Sure, Dr. Dajer," they nodded.

"And try to relieve her symptoms," I added. "You know, try Kaopectate, Imodium, that sort of thing." The stool tests would take another day. Treating symptoms without a firm diagnosis is the stuff of medical nightmares, but we had no choice.

The next morning I found that Mrs. Velez had been moved to a bed upstairs. When I checked in on her, she said the diarrhea had eased a bit. But despite lots of IV fluids, she looked haggard.

"I still haven't slept a wink," she said with a sigh. "Last night I felt like I was going to spend the rest of my life on the toilet!"

Out of fresh ideas, I offered a few words of consolation, then hurried on to the lab.

"Dr. Dajer!" one of the technicians greeted me. "We have a great slide for you!"

"Really? What?" I brightened. At least someone had made a diagnosis.

"*Strongyloides.* Tons of them. I've never seen so many on one slide."

"Huh," I bent down and fiddled with the knob on the microscope. "Who's the patient?"

"Velez."

"You're kidding, right?"

"No joke." The technician, normally a serious man, displayed a rare, brilliant smile.

Under the scope, dozens of larvae, coiled like tiny, translucent eels, zoomed into focus. They were a lucky find: stool specimens turn up the parasite in only about one-quarter of all *Strongyloides* cases. Sometimes the worm can only be detected through blood tests, biopsies, or probes of the intestine.

Strongyloides stercoralis has a name much longer than the twentieth-of-an-inch-long worm it describes. Most of the 80 million people afflicted with *Strongyloides* live in the tropical Third World, but Puerto Rico and the southern United States still lie within its reach. Yet Mrs. Velez hadn't lived in Puerto Rico for more than 40 years. And therein lies the first of the parasite's three formidable talents: its ability to reproduce and infect its host without ever leaving the body.

The worm's veritable fantastic voyage begins when filariform larvae—threadlike worms that live in the soil—burrow beneath a person's skin. After reaching the veins, they are carried to the lungs. From there they crawl up the windpipe and, with a wheeze or a cough, are soon swallowed down the esophagus. They then pass through the stomach and finally lodge in the small intestine, where they mature and lay eggs. The eggs hatch into more larvae that are shed in feces to start a new reservoir of worms in the soil.

Lots of parasites do that: mature in the host to produce eggs or larvae that are excreted in the feces. What makes *Strongyloides* exceptionally cunning is that it can become an infectious larva without ever leaving the host's gut. Once it has reached that stage, it can burrow out of the intestine into the bloodstream and begin a new cycle of infection, just as if it were penetrating the skin for the first time.

Strongyloides' second perfidious talent is its ability to cause symptoms far from the small intestine where it lodges. As it tunnels through the lungs on its way up the windpipe and down the esophagus, *Strongyloides* provokes an inflammatory response that mimics asthma.

But here's the coup de grâce: when disease or malnutrition weakens a host's immune system, many more larvae can make their way out of the intestine and up through the lungs. They eventually settle back in the intestine, dramatically boosting a patient's "worm load." The results are fever, severe abdominal pain, and diarrhea—the very symptoms that brought Mrs. Velez to the ER.

And perversely, nothing allows the worms to flourish better, than the steroids used to control the asthmatic symptoms that *Strongyloides* triggers. By hampering normal immune function, Mrs. Velez's repeated courses of steroids over two decades had slowly allowed *Strongyloides* to reach a critical mass her intestine.

Strongyloides was likely to be the true foe in Mrs. Velez's long battle with asthmatic wheezing and bronchitis. But we wouldn't know for sure until we rid her of the parasite. Luckily, there is strong medicine—thiabendazole—for the worm. Wily though it is, *Strongyloides*, too, would fall before the armamentarium of modem American medicine. The next day I expected to find a grateful patient well on her way to health. Instead, Mrs. Velez seemed about to leave tire treads on my shirt.

"I want to leave! Now! That little resident told me I was dirty! Here he is, a Chinese, telling me, a New Yorker, I must have come from a poor, dirty country to get this disease. And then they shut me in! As if I were contaminated. I'm leaving. Get me the papers!"

I coaxed her back into an armchair.

"But what happened?" I stammered.

Apparently, Mrs. Velez's doctors had approached her with all the finesse of a search-and-destroy mission. True, the same larvae that can reinfect the patient within the intestine can, if strict hygiene is not observed, be spread from feces to hands or sheets and then to other people. But the team of residents didn't bother to explain this to Mrs. Velez. Instead, they had simply clapped her in isolation and refused to touch her without gown and gloves.

"*Doñita*," I pleaded, "they're only interns ...." This was a half-lie. "They thought they were doing the right thing. But what's most important is that you stay for another day. The treatment takes two days. If you don't, you will only get sick again."

After much cajoling, she agreed to stay. But *Strongyloides* is a stubborn adversary. Because the eggs and larvae can survive treatment that kills the mature worms, many patients are not cured after just one course of thiabendazole. To be sure the parasite was eradicated, Mrs. Velez needed to come back for more stool tests and possibly more medicine.

She never did.

Strongyloides had triumphed again by triggering precisely the wrong response in its foes. By provoking too vigorous an immune response in the lungs, the parasite causes asthmalike symptoms. When doctors attack the "asthma" with steroids, the wily worm runs rampant. So, too, in Mrs. Velez's case, the presence of *Strongyloides* set off a prejudice in her doctors' minds that she came from a "poor, dirty country." By treating her as if she were somehow unclean, her "modern" doctors drove Mrs. Velez away and gave *Strongyloides* another lease on life.

UPDATE

Strongyloides stercoralis *is endemic in tropical and subtropical countries, such as West Africa, the Caribbean, and Southeast Asia, where prevalence rates can be as high as 40 percent. The disease is estimated to affect more than 70 million people worldwide. Long infective periods, as described in this encounter with* S. stercoralis, *are not uncommon. In 2004, a 49-year-old woman was admitted to a Netherlands hospital after a three-week history of colitis-like symptoms of vomiting, diarrhea, abdominal cramps, and fever. The patient had been born and raised in Suriname, South America, and immigrated to the Netherlands in 1977 at the age of 22. Netherlands doctors subjected the woman to numerous tests before they discovered parasitic worm larvae in her feces. Since* S. stercoralis *is not found in the Netherlands, the woman's doctors believe she must have been infected in Suriname 27 years before. She was successfully treated with ivermectin. Today, ivermectin is the drug of first choice for treatment because thiabendazole, the drug administered in the above story, is no longer recommended due to its side effects and lower efficacy. Even with ivermectin though, repeated treatments must be administered to ensure all adult worms, which develop from the autoinfective larvae, are killed.*

QUESTIONS TO CONSIDER

1. Was the delay in determining a correct diagnosis detrimental to Mrs. Velez?

2. What was the sequence of disease possibilities that were considered before the correct diagnosis was made?

3. How did a diagnosis of asthma exacerbate Mrs. Velez's disease?

4. How did the symptoms displayed by Mrs. Velez relate to the final diagnosis?

5. Describe some of the characteristics of the infectious agent causing Mrs. Velez's disease.

When we think of heart disease, we usually conjure images of fat accumulation, clogged arteries, leaky valves, and other physiological aberrations of the body. Rarely do we think of infectious disease, but sometimes a microorganism can be responsible for a debilitating heart illness, as this encounter relates.

INTRUDER IN THE HEART

CLAIRE PANOSIAN

Miguel never recalled when they actually began—those feeble, quivering heartbeats that snaked through his chest like a high-speed train. The first attacks were so brief he forgot them as soon as they passed. Then, over time, they came more often and lasted longer, until one day a friend at work saw the whole crazy business, the sweat on the face and the fear in his eyes. He thought Miguel was having some kind of dizzy spell and made him lie down. Soon after, Miguel's wife drove up in their truck and took him to the emergency room at L.A. County Hospital. That's when they did an electrocardiogram, his first ever. "There are extra heartbeats," said the nurse, "but no sign of heart attack." The intern on duty added, "You're 38 years old, your blood pressure's great, you don't even smoke. You've got a lot of miles left on that heart." So Miguel decided it was nothing. From then on he pretended he was fine.

The following winter Miguel caught a cold from his daughter. The sore throat and aching muscles kept him in bed. Soon his feet and ankles began to swell, and he couldn't get air. Finally he spent an entire night bolt upright (since he could no longer catch his breath lying down), wheezing, coughing, and gurgling like a worn-out radiator. That night he also worried. He could no longer fool himself. He could tell that something was terribly wrong.

I'm an infectious-diseases specialist in Los Angeles, and I first met Miguel while doing rounds with students and residents in our hospital's Coronary Observation Unit. That's where "medium sick" heart patients are sent, as opposed to the sleek Coronary Care Unit, which is filled with sufferers of crushing chest pain and out-and-out cardiac disasters. I guess Miguel was only a slow-motion disaster. Over a period of months, he had gone from clinic to clinic. Finally he received a heart catheterization. In this imaging procedure, dye is squirted through a tube placed in the heart to reveal its structure and function. The results showed no coronary blockage. After that, Miguel carried the diagnosis of congestive cardiomyopathy and ventricular tachyarrhythmia, idiopathic. Basically, that meant he had a flabby balloon of a heart beating dangerously out of control, and no one knew why. His heart and circulatory system were functioning so poorly that fluids were building up. He had come to us for a possible heart transplant.

Before the students and I visited him, I scanned Miguel's chart. Since his first attack of waterlogged lungs, he had taken digitalis and diuretics. The digitalis helped his heart pump more efficiently, and the diuretics helped flush out the fluid that was building up from the failing pump. He also stopped working—"Too tired and short of breath," read the social worker's note. On his current chest X-ray, his heart was a large baggy mass in the middle of his chest. But at least his lungs were clear; thanks to more potent diuretics, in the last few days they had been wrung free of fluid like a wet mop.

We left the nursing station and walked to Miguel's room. Seated in bed, oxygen tubing dangling from his neck, he was speaking Spanish on the phone and looked comfortable enough. When he caught sight of us, he hung up. Then I saw his face. "My God," I thought, "he's my age, but what old eyes." They were dark, deep pools with just a glimmer of hope. It was the hope that scared me.

With the help of a Spanish-speaking student, I launched into my polite prologue: who I was, what I did, who sent me. The usual formalities.

Miguel replied, *"Señora,* thank you for your time, but are you sure you're in the right room? You're the fifth doctor today, and not even a heart specialist. What's this about infection? I'm clean. Faithful, too. Not that it matters much, the way I feel now."

Thus began the intricate dance between doctor and patient. Language was the least of our barriers. My arrival made no sense to Miguel, and why should it? Even to many M.D.'s, microbiology is a far stretch from heart disease. True, we're all taught that viral infections can produce heart inflammation. Often, however, cardiomyopathy patients' damaged muscle fibers are eventually blamed on clogged arteries or too much booze.

The puzzle in Miguel's case was that none of these theories fit. Not that he cared. As he lay slowly dying, he wanted a mechanical fix, not a medical lecture.

Mechanic I'm not. What I am is a parasite sleuth. And today my quarry was *Trypanosoma cruzi*, a single-celled protozoan that can burrow into heart muscle. An ancient, sickle-shaped organism indigenous to Central and South America, it remains in the 1990s one of the leading causes of cardiac death in young and middle-aged people south of the border. Right now, the invader inhabits 16 to 18 million humans in Latin America and more than 300,000 U.S. immigrants. But most victims are unaware of their infection, as are their doctors. Hence, even in southern California, *T. cruzi* infections are missed more often than they are diagnosed. I didn't want to miss the diagnosis in Miguel. And so I pressed for answers.

"*Doctora*, these questions about my home—what does it matter?" Miguel asked. "All right, Jalisco was the place. My family has a small farm with goats and pigs and chickens. Our house? Oh, adobe—mud and straw—very poor. You'll find no fancy homes there. And yes, there are bugs in the house. Naturally. They have to live, too."

Slowly the clues were adding up: farm, animals, insects. The natural reservoir of *T. cruzi* is warm-blooded animals— lots of animals, currently numbering more than 100 species. House-dwelling insects, specifically known as conenose, or "kissing," bugs (entomologists call them reduviids) transfer *T. cruzi* from animals to man. The final clue was poverty. That's because reduviids often nest in the nooks and crannies of the poorest rural dwellings, wattle and daub houses. From this retreat, the bugs emerge at night to binge on humans and purge: they first suck blood, then dump parasite-laden feces next to the skin punctures they've just created. Most victims sleep through the insult.

Carlos Chagas was the first scientist to expose the secret life of *T. cruzi*. Sent to central Brazil in the early 1900s to battle malaria, he spent his spare time viewing the intestinal contents of reduviids through his microscope. What he found were crescents that looked like *Trypanosoma brucei*, the then recently discovered cause of African sleeping sickness. Chagas later saw similar forms in the blood of other mammals, local residents, and nearby railway workers. He suspected the parasite was linked to the heart malfunction among the infected. Though the pathogen's taxonomic name, *Trypanosoma cruzi*, pays homage to Chagas's mentor, Dr. Oswaldo Cruz, most doctors today know the infection as Chagas' disease.

By now we were ready to examine Miguel. The team watched as I bent over his chest with my stethoscope. I was ready to hear the soft syncopation of a failing heart, but when I closed my eyes to focus, another image came to mind. I envisioned the moment years ago when the microscopic creatures entered Miguel's body and were swept through his blood, settling into tissues and building cystlike homes. Over the years, they reproduced in his brain, nerves, skeletal muscles, and heart. I pictured the protracted war they waged with his body's immune system, leaving heart and other tissue strewn with dead and dying cells. I tried to explain to him that this simple one-celled organism gave 25 to 30 percent of its victims chronic heart problems just like his.

Miguel did not understand. *"Doctora,* please excuse me, but you must be mistaken. I have been to hospitals many times in the past year. If I had *animales* in my heart, don't you think they would have known by now?"

Trusting Miguel. I would not burden him with studies that told otherwise. But later that day I resolved to show my team an article published in 1991 in the *New England Journal of Medicine*. In it, cardiologists at the biggest public hospital in Los Angeles hypothesized that a few patients with Chagas' disease could easily be lost among the far larger tribe of atherosclerosis sufferers patronizing county clinics. They proved their theory by finding 25 patients with Chagas' on their rolls who had previously been labeled" coronary artery disease" or "dilated cardiomyopathy" with unknown causes. Some of these *T. cruzi* victims had carried the misdiagnosis up to nine years.

I turned to Miguel's abdomen. First I felt for his liver. It was easy to find, a fleshy peninsula three inches below the ribs. It was slightly enlarged, confirming a blood buildup from an ailing heart. As my fingertips continued to search, they also sensed fullness of the colon. Did he have constipation, I inquired through the student. No, the answer came back, not usually.

I was relieved. In my mental cinema, I had imagined single-celled squatters, this time ensconced in the nerve clusters that lace around the digestive tract. Parasite nests and inflammation located here produce "mega syndrome," Chagas' disease's strangest hallmark, seen in 6 to 10 percent of its victims. The gut enlarges because the nerves that control the rhythmic squeezing of peristalsis die. Such nerve damage can afflict other hollow organs too. Autopsies on mega patients play like bad horror flicks. With the knife's first stroke, the overstretched organ spills out like a membranous balloon. On top of heart disease, this was almost too cruel to contemplate. Thank goodness Miguel had been spared the indignity.

Now Miguel spoke up, and for a moment, we glimpsed his private storehouse of thoughts and fears.

"*Señora*, if you are right, could I give this infection to my family?"

I was relieved that on this point, at least, I could reassure him. Except for occasional cases of maternal-fetal transmission, intrafamilial spread of *T. cruzi* does not occur, except by the feces of reduviids. But blood donation is a different story. In Santiago, Chile, the prevalence of *T. cruzi* -positive blood is 2.6 percent, and in Buenos Aires, Argentina, 4.9 percent. These and other Latin American countries have made screening blood for Chagas' disease compulsory by law. Not so the United States. Admittedly, the prevalence of antibody in blood donors is lower here. But it was recently pegged at 1 in 8,800 blood units in both Los Angeles and Miami. Considering that anywhere from 13 to 23 percent of those antibody-positive units harbor enough organisms to transmit Chagas' disease, a lot of folks in my field find our lack of testing wholly unacceptable.

But pondering the transfusion issue was not our task. The time had come to leave Miguel's bedside and answer the fundamental question: Did he have Chagas' disease?

Our best bet for starters was the antibody test for *T. cruzi* performed at the Centers for Disease Control and Prevention in Atlanta—it is simple, clean, and fast. I couldn't imagine describing the alternative to Miguel. A few years earlier I had been one of several doctors huddled around a Peruvian woman about to undergo xenodiagnosis, once considered the gold standard for diagnosing chronic Chagas' disease. Understandably, the woman was queasy. Xenodiagnosis requires roughly three dozen hungry, laboratory-reared reduviid bugs to serve as biological incubators. I recalled the shades being drawn, the hospital lights dimmed. Then the insects were allowed to probe and siphon from our patient's forearms. After drinking their fill and defecating, the hostages were returned to their cages. Weeks later, their diminutive guts were dissected and searched for *T. cruzi* parasites.

I never had to discuss xenodiagnosis with Miguel. One week after drawing his blood, I heard from a technician in Atlanta. Miguel's serum was unequivocally positive. Then the technician posed a question. Since Miguel was from Mexico, a country thought by some to have less virulent strains of *T. cruzi*, how sick was he, really?

Everyone in medicine has strengths and weaknesses. As a clinical specialist in tropical medicine, I do not pretend to know about regional differences in parasites. But I do know patients. No, Miguel would not die tomorrow. But, I added, a 40-yearold day laborer with end-stage Chagas' cardiomyopathy facing possible heart transplantation was, by any criterion, sick. After hanging up the phone, I imagined the conversation I would like to have had with Miguel.

My friend, it's as we thought: unknown to you, a migrant and his family have been living in your heart. Years ago, these creatures came looking for a home for their descendants, and in you they found the perfect place. The good news is this: we have a strong medicine. With a few doses the trespassers will be evicted, and you'll be as good as new. Back to work, back to your wife, back to Sunday *menudo* with all the salt you like.

That was my dream. Sadly, it was far from reality. In truth, medical treatment had never been the issue. The cardiologists who asked me to see Miguel suspected Chagas' disease long before I appeared. They also knew there was no magic bullet for this stage of *T. cruzi* infection. There are drugs to treat Chagas' disease, but they are toxic and not very effective and even less so when the infection has progressed. At this stage, the real question facing Miguel was heart transplantation. If Miguel had a transplant, he would need to take drugs to rein in his immune response and prevent rejection of his new heart. What would that do to the delicate balance of power between parasite and host? Would a transplant cause more harm than good?

I could tell him that patients with end-stage Chagas' disease had gotten new hearts—as of a few years ago, in Brazil there were 31 recipients, to be exact. But early results of transplantation had been dismal. Most patients died within a year from new colonies of *T. cruzi* sprouting not only in their new heart but in their skin, brain, and other organs. The drugs used to prevent organ rejection created an open playing field for parasite growth.

In the end, Miguel made his own decision. When he heard all the facts, he declined to be placed on the transplant list, and that was that. I didn't see him again. Months later, I heard he had died. He was one of more than 40,000 who lost their lives that year to a hidden intruder in the heart.

UPDATE

The Brazilian physician Carlos Chagas discovered human American trypanosomiasis (Chagas' disease) in 1909. His work is remarkable because he discovered the parasite in the reduviid bugs before describing all the epidemiological and clinical aspects of the infection. In 2007, an estimated 16 million to 18 million people are infected with T. cruzi *in 18 countries of Latin America. Of the almost 7,000 deaths annually in Latin America from Chagas' disease, about 75 percent occur in Brazil. Until recently, Chagas' disease in the United States has been rare. In late 2007, however, the Centers for Disease Control and Prevention (CDC) journal* Emerging Infectious Diseases *reported the first "home grown" case, and at the 2007 annual meeting of the American Society of Tropical Medicine and Hygiene, scientists reported in a study that one in 30,000 blood donors tested positive for* T. cruzi. *These donors were primarily people who had spent significant time in or emigrated from Latin America. Although most will not develop the disease, an international association involved in activities related to blood transfusion reported more than 317 people in 30 states were confirmed as having* T. cruzi *infections in 2006. The fear is that this might be just the proverbial "tip of the iceberg." Therapy currently consists of nifurtimox and benznidazole for acute cases; as this encounter described, there still is no effective therapy for chronic cases.*

QUESTIONS TO CONSIDER

1. How might transmission of the infectious agent to Miguel have been prevented?

2. Why would penicillin be useless in treating Miguel?

3. Besides reduviid bugs, what is the most likely way that the infectious agent in this encounter could be transmitted from one person to another?

4. Would a bite from a reduviid bug necessarily mean that an individual will develop illness as discussed in this encounter?

5. Why would most heart transplants associated with Miguel's disease result in failure?

Hope can spring eternal. While a patient battles on, the physician is encouraged to search for a successful resolution to the patient's illness. Even when the patient seems to succumb to despair, the physician must continue the search and never give up hope. In this encounter, we meet a woman whose illness would test the strongest of hearts. We see that even though she is ready to cave in to her illness, the physician does his best to lighten her load and bring her back from the brink of despair.

A STAR OF HOPE

PAUL ARONOWITZ

"Come to Room 1052 immediately," my voice pager squawked. It was 5 A.M., I was in the third month of my internship, and I had just finished working up an admission. I rushed two floors up and jogged down the hallway. The lights were dimmed for the night, and a chill wrapped around me as I hurried to Doris's room.

Doris was my favorite patient. A tragic testament to the ravages of systemic lupus erythematosus, she was a 42-year-old mother who had been virtually bedridden for nearly three years from the complications of her disease. In this devastating immune system disorder, the body attacks itself, creating inflammation in the organs, joints, and tissues. Doris had become disabled because of a serious infection in one of her hips. The hip joint had been removed, but it was never replaced because she was never stable enough for another operation.

Her kidneys functioned so poorly that she was close to requiring dialysis. She had anemia, a stomach ulcer, and chronic diarrhea. Like many lupus patients, Doris was being treated with steroid hormones that suppressed the immune cells that were causing her problems. But the steroids that suppressed Doris's lupus also weakened her defenses against bacteria and viruses. She had suffered from pneumonia and urinary tract infections. Almost every organ in her body was directly or indirectly under assault from her disease.

At the time I was caring for her, Doris was being treated for a large infected ulcer on her lower back. Because of life-threatening infections from the bacteria that had colonized her ulcer, she was getting intravenous antibiotics, and twice in the previous two weeks her intravenous lines had become infected.

Some doctors might have found Doris's case depressing, the inevitable progression of her lupus an empty and demoralizing vision of how far we have yet to go in treating and curing chronic disease. I didn't happen to feel this way.

Though she seemed to develop a new medical problem every day I took care of her, I enjoyed Doris's company. Her laugh was infectious, her humor dry and self-deprecating. Despite her disease, despite being divorced and trying to raise a rambunctious 11-year-old daughter alone, Doris managed to hold on to her dignity and sense of humor. Somewhere deep in her heart she kept alive dreams of being well, an exploding star of hope that I was lucky enough to glimpse from time to time.

"Help me with this unruly daughter of mine," she would say. "Get me well so I can get up out of this wheelchair and show her who's boss." She would laugh loudly, her gnarled, arthritic hands crossed in her lap, tears filling her eyes, a broad smile across her face.

I had no doubt that Doris believed she would get well. After all, I imagined her thinking, no one her age with a daughter, an apartment, and a life to live could afford to remain sick for long. She had once worked as a clerk, but it had been years since she was able to type. Now her disfigured hands could barely hold a pencil to do the crossword puzzles she worked on late into the night.

To my astonishment, despite her severe illness, Doris smoked cigarettes. It seemed to be her way of saying that she could do just as many unhealthy things to herself as any healthy person. She regularly sneaked smokes in her room, drawing on the cigarette by the open window and flipping out the butt the moment someone knocked at her door. No one ever caught her, but we all recognized the smell of cigarette smoke that wafted periodically from under the door. She refused to discuss the issue. "You're my doctor," she'd say with a smile, "not my mother."

As I jogged down the hall, I knew the call from nursing meant trouble. It was not Doris's nature to have a minor problem at 5 A.M. When I got to her room, I found four nurses crouched over Doris, who lay facedown on the floor in a pool of blood. She was muttering something about her bad luck.

Doris, an inveterate night owl, had been playing solitaire and drinking a diet cherry soda when she dozed off and fell forward out of her chair, landing on her leg and face. We put her in bed, and I spent the next two hours meticulously suturing a large gash over her right eye and another on her right thigh. I knew Doris's skin was exceptionally fragile, since the steroids she took for her lupus also caused the protective barrier of collagen fibers beneath the skin to thin. I worried that the new wounds might become infected.

Over the next few days Doris's wounds slowly healed, and, to our surprise, things seemed to start going well for her. She found a temporary foster home for her daughter. The plastic surgeons were considering performing a skin graft that would cover the ulcer on her back and allow it to heal. And her lupus, for once, was not acting up.

But not long after that I stopped by Doris's room early one morning. I eyed her breakfast tray for traces of cigarette ash and noticed that her food was untouched.

"I just can't believe this," she said sadly. "It's as though someone out there doesn't want me to get out of this crazy place. I say my prayers, brush my teeth, and don't hurt anybody. I'm even a nice person sometimes." A tear rolled down her cheek.

"Doris, what are you talking about?"

"I'm talking about this!" she cried, raising both arms toward me. Her hands hung limply from her wrists. She couldn't lift them.

As I examined her, I was unable to reassure Doris that she would be able to lift her hands again. I was completely baffled about what could have caused the "wrist drop."

Most of the specialists in the hospital were already familiar with Doris's case, so a few quick phone calls drew a flurry of doctors to her room. Each came up with a long list of different reasons for Doris's problem. No two seemed to agree about what was wrong.

The rheumatologists, who treat lupus since it is an inflammation of the joints, thought the disease was now causing inflammation in her blood vessels and nervous system. In a booming, confident voice, a senior rheumatologist matter-of-factly advised me to give Doris higher doses of steroids and to call the surgeons in to do a biopsy of her sural nerve, a large nerve near the calf. Although the biopsy would involve incising her fragile skin, inspecting the nerve for inflammation might confirm his suspicions.

The neurologists thought the steroids were causing Doris's wrist drop and advised me to reduce or even discontinue them. They advised me to order nerve conduction studies, tests in which needles are inserted into opposite ends of the arm to measure the rate and quality of nerve signals. A disturbance in the signals might tell them what was going wrong.

The infectious-disease specialists suspected a new bacterial or viral agent was at work. They recommended taking bacterial and viral cultures of Doris's blood and giving her more antibiotics.

Late that afternoon I went to Doris's room and explained what the specialists had recommended. Doris sat staring down at her hands. So much of her body was useless to her now. I tried to imagine what she was feeling. I couldn't. She lifted an arm and rubbed the back of it across her eyes to wipe away the tears rolling down her cheeks.

"In other words," Doris said bluntly, "you guys have no idea what I have." I nodded. "And now you want to poke needles in my arms, connect me to AC current, cut my leg open to snip out pieces of my nerves, and raise or lower my steroids—you don't know which—until you come up with something?" She shook her head in disgust.

I began to defend the specialists, carefully repeating the reasons for the procedures, but I stopped midsentence. I realized that to them Doris was an interesting and frustrating case of lupus. She was not a complicated 42-year-old mother intent on walking out of the hospital with a cigarette tucked between her lips. To them she was the patient described in a stack of medical records that stood four and a half feet high.

"I'll have no part of it," she said after a few minutes of silence. I felt exhausted and overwhelmed by Doris's despair. It was early evening, and her small hospital room was cloaked in gloom. A thick San Francisco fog blew by outside. I realized I'd been on call for nearly 36 hours. I walked out, leaving Doris sitting by the window, watching the light fade.

When I stopped by the next morning, Doris said, "I'm sorry I was mean to you yesterday."

"You were angry. It doesn't matter."

"Are we still friends?" she asked.

"Of course."

"Gee, how I could use a cigarette," she said, with a soft chuckle. Doris couldn't lift her hands to light or smoke a cigarette. The nurses had to feed her meals because she could no longer hold a fork.

Later that day I met with my supervising attending physician. A quiet, methodical man, Dr. Rood listened carefully as I told him what had happened with Doris. Sensing my hopelessness, he came back with me to reexamine her.

"You know," he said to me as we left the room, "she can try and then give up, but you're her physician. You're not allowed to give up."

I was silent, stung by Dr. Rood's gentle reprimand.

"Go back to her chart," he said, putting his hand gently on my shoulder. "Go over it carefully. Make a list of her medicines and we'll scrutinize it together. We need to make sure that this problem—this wrist drop—is not something we're doing to her."

Doris was taking 25 different medications. I worked late into the evening reviewing her chart and taking notes, making the medicine list, and going over tests she'd had before I began caring for her.

The next day Dr. Rood and I met in his office. A breathtaking view of the Golden Gate Bridge filled the window behind his desk; the pages of a half-edited medical textbook lay open before him.

"I'm sorry," he began. "I didn't mean to say you aren't doing a good job of caring for your patient."

"You were right. I shouldn't have given up. It's just that—a"

"You're friends," he finished for me.

I remembered playing gin rummy with Doris at 3 A.M. on an unusually quiet call night several weeks before.

"No matter what happens with all of this," said Dr. Rood, "Doris will have a very hard time when you rotate off service in a few weeks." Coming from him, this was a great compliment.

We went over my notes, and then Dr. Rood sat back in his chair, rubbing his mustache.

"Why is Doris on metronidazole?" he asked, referring to an antibiotic that Doris had been taking for almost six weeks. Metronidazole, commonly known by the brand name Flagyl, is frequently used to treat infections caused by anaerobic bacteria, bacteria that don't require oxygen to grow. Doris was taking it for her infected back ulcer.

"I seem to recall," said Dr. Rood, "that metronidazole can occasionally cause numbness in an extremity. If the drug was interfering with Doris's nervous system, it could explain her wrist drop."

"Is it reversible if the drug is stopped?"

"I honestly don't know."

"I'll try anything."

"Then go to it," Dr. Rood said, shaking my hand.

I stopped by Doris's room to tell her that we were taking her off metronidazole and to explain why. She was sitting in a chair by the window, and her daughter was bouncing up and down on her hospital bed.

"Sounds like a shot in the dark, if you ask me," she said. "Well, I didn't," I joked, feeling hopeful again, "and it is." Doris chuckled softly.

There was no change the next day or the day after, but around midnight that night, I was almost caught up on my work, so I wandered over to Doris's room. Like some desert animal, she was most alert and active between midnight and 5 A.M. As I knocked at her door, the unmistakable odor of cigarette smoke greeted me. When I opened the door, I found Doris seated by the open window, grinning sheepishly, her hands empty.

"They're working again," she offered matter-of-factly.

"What?"

"My wrists. I can't do everything, but watch this." With great effort she lifted her hands from their drooping position and smiled proudly.

Over the next several days, with the help of physical therapy, Doris's wrists returned to normal. Her back ulcer had also improved, so that she no longer needed antibiotic treatment. She was in a good mood again.

"Listen," she said a few nights later, staring down at a hand of gin rummy. "I know I'm not the easiest person to take care of, but I want you to know that I appreciate all you've done."

I thought about how little I'd really done for her—she was still in the hospital, and her daughter was in a foster home because her mother couldn't take care of her anymore. Doris couldn't walk, work, be a mother, or enjoy the pleasures we so often take for granted. She was able only to dream of a time when things would be better.

Yet, in the end, I had taken care of Doris by attending to details like her wrists. These details, I knew, were not small things to her. Doris had taught me a lot about what it means to be a doctor. There was no way to tell her that she had become more than a patient—that in her own quiet way she had become a friend and, finally, a teacher.

"Don't mention it," I said. And I drew another card from the deck lying facedown on the table between us.

UPDATE

The immune system of someone suffering from lupus attacks healthy cells and tissues by mistake—such disorders are referred to as autoimmune disorders. There are many forms of the disorder. Systemic lupus erythematosus affects many parts of the body, including the joints, skin, blood vessels, and organs. As described in this encounter, if the individual's disorder is not well controlled, it can result in significant disability. Anyone can be stricken with lupus, but women are most at risk. Although the cause of lupus has remained unclear, in 2007 researchers identified a genetic variation that increases the risk of lupus. The genetic profiles of patients and controls in the study seem to indicate that individuals carrying two copies of the gene variant have more than double the risk for lupus compared with people who carry no copies of the variant form. It remains to be determined how the gene variant increases the risk for developing lupus.

QUESTIONS TO CONSIDER

1. What sort of disorder is lupus, the illness affecting Doris?

2. What are the complications of Doris's disease?

3. How did the different specialists suggest treating Doris?

4. Suppose the infectious disease specialist had been correct in his diagnosis of a viral infection. Would the administration of more antibiotics have helped? Explain.

5. What is the major lesson that should be apparent from this encounter?

Medicine is both an art and a science. Technology supports the science of medicine, but instinct supports the art. When the technology is not available for a diagnosis, the physician must trust instinct. In this encounter, a physician pays heed to his gut feeling and treats the patient accordingly. If he is wrong, disaster awaits.

AN INDEPENDENT DIAGNOSIS

TONY MIKSANEK

Caitlin literally stumbled into my office late one sweltering afternoon at the beginning of July. She didn't have an appointment, but that didn't matter: she was swaying on her feet, gripping her head in her hands, and looking around somewhat blankly. I knew she had to be seen right away. After my nurse and I helped her onto the examining table, she curled her body into a ball of pure misery. With just about any patient, such a posture would be alarming, but with Caitlin it was nothing short of shocking. I have a family practice in a small town in Illinois, and I'd known Caitlin for much of her 23-year life. Having seen her through fevers, sore throats, and even an appendicitis attack that she had laughingly described as "just a bellyache," I knew she had a pain threshold unrivaled by most.

"My head feels like it's ready to explode," she said huskily, wincing. She looked up, her hands still clamped around her head as if to keep it from doing just that. "At times I feel like I'm burning up, then suddenly I start shaking with the chills. I feel like I've been run over." Her voice dropped to a whisper and she started to cry. "Help me."

To do that, I first needed to get her recent medical history—what she'd been up to, where she'd gone, who she'd seen. It's the most important information a physician can have; an astute listener with the right questions and a modicum of patience can often make a diagnosis from a history alone. I learned that Caitlin had returned just a week earlier from a vacation in Florida, perfectly healthy and in great spirits. But for the last two days she'd been fighting the increasingly excruciating headache and the fever and chills she'd already described; in addition, she revealed, she was nauseated, the muscles in her back and legs ached, she had a sore throat, and she was exceptionally weak.

As we talked I started compiling a list of possible diagnoses in my head, anything that would explain her symptoms and their severity. Could it be meningitis, an inflammation of the membranes covering the brain and spinal cord? She didn't think she'd been exposed to anyone with the disease, but she couldn't be sure. Might it be a severe flu or other viral infection? What about mononucleosis? She couldn't think of any friends or family members with similar symptoms, but again, she wasn't positive. Food poisoning? She didn't think she'd ingested any suspect food or tainted water, but after all, she had been on vacation. Lyme disease? Rocky Mountain spotted fever? She hadn't seen any ticks on her body, or any tick bites, but she'd done some camping in Florida and might just have overlooked a tick. Toxic shock syndrome? Well, she had just completed her menstrual period, and she did use tampons, so that was also a possibility.

There wasn't much I could do for Caitlin in my office: I decided to get her admitted to our small local hospital. I knew I had made the right decision when she didn't even argue with me.

When I got a chance to thoroughly examine Caitlin, I found she had a temperature of 104 degrees and was slightly dehydrated and quite weak. Her eyes were mildly sensitive to light, which can be a sign of meningitis, but her neck wasn't stiff—an argument against that diagnosis. The nurses took samples of her blood, urine, and sputum (the stuff that comes up when you clear your throat) and sent them to the lab, asking the technicians to make sure to test the blood for mono, Lyme disease, and Rocky Mountain spotted fever. I ordered a chest X-ray to rule out the possibility of pneumonia or some other respiratory ailment. And before the night was out, I performed a spinal tap in hopes of ruling out meningitis.

The spinal fluid showed no signs of meningitis. The chest X-ray was clean. The mono test was negative. A blood count showed that Caitlin had a surprisingly normal number of white blood cells—the number should have been at least twice as high in someone with an infection as severe as Caitlin's appeared to be. A few of the results were slightly off: she had a somewhat low level of sodium and a borderline-low protein level in her blood, and her liver appeared to be working a little bit harder than usual. Unfortunately, these are nonspecific findings. They're pieces of the puzzle but they don't fit into any real pattern. They're the kinds of signs you can't pin a diagnosis on.

So, on paper, the case was still muddled. But although some doctors would never admit it, and patients might be frightened to hear it, intuition plays an important role in the practice of medicine. And my intuition was suggesting to me that Caitlin had Rocky Mountain spotted fever. The blood tests would take a while to come back, but in the meantime I'd already eliminated meningitis and pneumonia. Her blood count and other tests made it seem unlikely that she had a viral infection or toxic shock syndrome. Equally important, though, they didn't rule out Rocky Mountain spotted fever: for unknown reasons, the white blood cells don't initially respond to this infection the way they do to most others, so early blood counts are often normal in these patients. And though Caitlin didn't yet have the characteristic rash found in more than three-quarters of spotted fever cases and she didn't remember a tick bite, the rest of her symptoms fit—headache, fever, nausea, muscle aches, lethargy.

Rocky Mountain spotted fever is an infection caused by the bacterium *Rickettsia rickettsii*, which grows inside the cells of ticks and mammals. Ticks can pick up the bug by feasting on the blood of an already infected rabbit or rodent, and they can pass it on to their offspring. The disease was so named because Rocky Mountain spotted fever was first reported in states such as Montana and Idaho. Ironically, it is now predominantly an Appalachian disease: the majority of cases are found in the Carolinas, Georgia, Maryland, Virginia, and Tennessee, though it can be found almost anywhere in the country (a large number of cases are reported in Oklahoma, for instance). The fever itself comes from the bite of an infected tick—wood ticks in western states, dog ticks in southeastern states—or from crushing one of these ticks on skin that has already been cut or pierced. (That's why you're supposed to use tweezers to pull off any ticks that attach themselves to you and not just mash them with your thumb.) As often as 20 percent of the time, however, the patient is either unaware of a recent tick bite or has forgotten about it.

Once in the bloodstream, the Rickettsia bacterium sets off widespread reactions. The immune system's response to the invader and the toxins given off by the bugs causes inflammation in and damage to blood vessels throughout the body. This can lead to a whole host of complications: encephalitis, pneumonia, kidney failure, shock. If it goes untreated, Rocky Mountain spotted fever can kill as many as 20 percent of the people it strikes; even when treated, 6 to 7 percent of patients still succumb.

That's why my hunch didn't stop me from worrying about Caitlin. And I'll admit I worried a little bit more when the sensitive blood tests for antibodies to Lyme disease and Rocky Mountain spotted fever came back—both were negative.

This sort of situation emphasizes the occasional conflict between the art and science of medicine, the battle between instinct and technology. In my gut, I knew what Caitlin had. But in my hands were test results telling me something else. Keeping in mind that it can take as many as four weeks for patients with this infection to develop the antibodies these tests look for, I decided to go with my gut.

I had already started Caitlin on two intravenous antibiotics. One was a drug that penetrates into most body fluids and tissues and is effective against a broad spectrum of bacterial infections. The other was doxycycline, an antibiotic that has a good track record in treating disease caused by unusual bugs like Rickettsia and the Lyme bug, *Borrelia burgdorferi*. In addition, I ordered IV fluids, painkillers, and acetaminophen for her fever. Then there was nothing to do but watch and wait. Patience is a virtue for both doctors and patients, though arguably easier for the former.

On her second hospital day, Caitlin's condition worsened. She continued to spike high fevers and she became increasingly confused.

"Are you feeling any better?" I asked her at one point.

"Huh?" was her only answer. Repeating the question didn't help. Eventually I did get her to answer a few simple questions—she told me her name and knew that the woman by her bed was her mother—but she couldn't tell me what day of the week it was or even the month. She was frighteningly weak for such a young woman, despite the care of family and friends constantly huddled around her bed. They reminded me of covered wagons, circling around to defend against an enemy. But Caitlin was almost oblivious to their presence, and I was unable to satisfy them with my answers to their questions. They seemed to expect me to pull magic out of my black bag when all I could offer them was hope.

On her third hospital day—the Fourth of July, as it happened—Caitlin's hands and feet became puffy. Even though she still wasn't responding to the antibiotics, I was encouraged. This was the sign I was looking for. By that evening, a faint pink rash had appeared on her wrists and ankles—a very fine rash under the skin, the kind of rash you can see but can't feel. A Rocky Mountain spotted fever rash. So on July Fourth, Caitlin and I celebrated—not independence but diagnosis.

By the next morning the rash had become a darker shade of red, and it began to move from her hands and feet to her arms and legs and then to her trunk. This unique rash, and its unique pattern of spreading, is the distinguishing feature of this infection. Only occasionally does it spread to the face and only rarely does it itch. But because the rash is the result of the inflammation of tiny blood vessels under the skin, it can diminish blood supply to the area and lead to gangrene.

Now that I had my diagnosis, I changed Caitlin's medication slightly, continuing the doxycycline but stopping the other antibiotic. And because she was still so ill, I added intravenous corticosteroids to her treatment because high doses of these drugs, naturally produced in smaller quantities by the body, decrease inflammation and help keep the blood circulating and blood pressure steady. Not everyone agrees that corticosteroids should be used to treat Rocky Mountain spotted fever, but it was the last card I had to play; I hoped it would hasten her recovery. It can take a long time to recover from Rocky Mountain spotted fever, because while the antibiotics can stop the bacteria from reproducing, they can't get rid of them altogether. That takes the immune system, and the immune system can take its time.

Indeed, it took a few more days, but Caitlin slowly improved: her temperature started to come down and her rash began to fade. After 48 feverless hours, I took her off the IV antibiotic and fluids, substituting doxycycline in pill form. Freed from the restraint of the IV line and with her energy level starting to rise again, Caitlin was like an uncaged bird. The morning I walked in as she was doing her hair and putting on makeup was the morning I realized she was well enough to go home. I practically had to chase after her to give her discharge instructions and a prescription for her antibiotic.

It was two weeks later, at a scheduled follow-up visit, that her blood test finally came back positive, confirming that she'd had Rocky Mountain spotted fever. But her appearance and attitude had already told me all I needed to know.

"That must have been some high-octane medicine you used to cure me," Caitlin laughed as she strolled out of my office, steady on her feet.

"It certainly was," I politely agreed. But I knew that even though it was the antibiotic that hastened the cure, it was the resilience of youth, the timeliness of a peculiar rash, and the stubbornness of a country doctor that did the healing.

UPDATE

The longer a tick infected with Rickettsia rickettsii *stays attached to the skin, the greater the chance of developing Rocky Mountain spotted fever (RMSF); usually several hours of attachment are necessary. RMSF is not contagious from person to person. RMSF is the most severe and most frequently reported rickettsial illness in the United States. Although the Centers for Disease Control and Prevention (CDC) typically report about 250 to 1,200 cases of RMSF each year in the United States, a record 1,936 cases were reported in 2005. The best way to prevent tickborne diseases is to avoid being bitten by ticks. Use tick repellents that contain DEET, and wear long-sleeved shirts and long pants to prevent ticks from getting onto the skin. Check your entire body for ticks after being in any tick-infested areas. If you find attached ticks, they should be removed promptly. Remove the ticks with fine tweezers—grab the tick firmly by the head or as close to the head as possible and pull. Medical experts suggest you do not use heat, petroleum jelly, or other methods to try to make the tick back out on its own.*

QUESTIONS TO CONSIDER

1. What are some of the characteristics of the bacterial species infecting Caitlin?

2. Why did Caitlin's physician celebrate when her hands and feet became puffy?

3. What are possible complications of Caitlin's disease?

4. Is Rocky Mountain spotted fever an appropriate geographical name for the occurrence of this disease? Explain.

5. Which of Caitlin's activities probably exposed her to the pathogenic bacterium that infected her?

Glossary

Abscess A pus-filled cavity resulting from inflammation and usually caused by bacterial infection.

AIDS (acquired immunodeficiency syndrome) A disease of the immune system, caused by infection with the retrovirus HIV, which destroys certain immune cells and is transmitted through blood or bodily secretions such as semen.

Amebiasis An infection or disease caused by a protozoal parasite called *Entamoeba histolytica* that affects the intestines.

Amniocentesis A test to determine the health, sex, or genetic make up of a fetus by taking a sample of amniotic fluid through a needle inserted into the womb of the mother.

Anatomical diagnosis The identification of an illness or disorder in a patient by locating the physical site of symptoms.

Anemia A blood condition in which there are too few red blood cells or the red blood cells are deficient in hemoglobin, resulting in poor health.

Antibiotic A chemical substance, derived from bacteria or fungi, that can kill or inhibit the growth of bacteria.

Antibody A protein produced by certain immune cells in the body in response to the presence of a foreign substance.

Antiemetic Something that acts to prevent vomiting.

Antiretroviral drug An antiviral preparation used to treat HIV and AIDS.

Antitoxin An antibody produced in response to a particular toxin.

Arachnoid The middle of the three membranes that cover the brain and spinal cord.

Autoimmune disorder A malfunctioning of the immune system caused by the reaction of a person's own antibodies to substances naturally found in the body.

Blackwater fever *See* **Malaria.**

Blood-gas analysis A set of hematology tests that are performed to determine the concentration of oxygen and carbon dioxide gases, as well as bicarbonate and pH, in the blood.

Botox® A preparation of the botulinum toxin.

Botulism A serious paralytic illness, usually developing from food poisoning caused by eating preserved food that has been contaminated with the botulinum toxin.

Bronchitis An inflammation of the mucous membrane in the airways of the lungs, resulting from infection or irritation, and causing breathing problems and severe coughing.

Cardiomyopathy A disease of the heart muscle, usually long lasting and with an unknown or obscure cause.

Catheter A thin flexible tube that is inserted into a part of the body to inject or drain away fluid, or to keep a passage open.

CD4 count The number of immune cells (T cells) possessing CD4 receptors per cubic millimeter (mm^3) or milliliter (ml) of blood; used as an indicator to help physicians decide when to begin treatment in HIV-infected patients.

Cellulitis An infection and inflammation of the tissues beneath the skin.

Cesarean section An operation to deliver a baby by cutting through the mother's abdomen and womb.

Chagas' disease A sometimes fatal disease, occurring in South and Central America, that affects the heart and nervous system, and is caused by the protozoal parasite *Trypanosoma cruzi* and transmitted by blood-sucking insects (reduviid bugs).

Chloroquine A synthetic antimicrobial drug that is taken orally and is used to treat malaria.

Cholecystitis An inflammation of the gallbladder, usually caused by a bacterial infection or gallstones.

CSF (cerebrospinal fluid) The colorless fluid in and around the brain and spinal cord that absorbs shock and maintains uniform pressure.

CT scan A radiological scan in which cross-sectional images within a part of the body are formed using computerized techniques and are shown on a computer screen.

Cyst The protective covering of a protozoal parasite.

Cytokine A signaling protein secreted by many cell types that affects the activity of other cells nearby or quite distant; important in controlling an inflammatory response.

Decision tree A flow diagram or a logical sequence of steps for solving a problem (algorithm) used in making diagnostic decisions and for treating a patient.

DEET The active chemical ingredient in many insect repellents that are applied to the skin.

Diagnosis The identification of an illness or disorder in a patient through an interview, physical examination, medical tests, and other procedures.

Differential diagnosis A systematic method used by clinicians to identify the disease causing a patient's symptoms.

Dura mater The tough outermost membrane of the three that cover the brain and spinal cord.

Dystonia A neurological disorder in which sustained muscle contractions cause twisting and repetitive movements or abnormal postures; caused by physical trauma, an infection, or a reaction to certain prescription drugs.

Ehrlichiosis A tickborne disease that causes fever, reduced blood cell count, and extreme sensitivity to light.

Electrocardiogram A visual record of the electrical activity of the heart.

Electrolytes Ions [sodium (Na^+), potassium (K^+), calcium (Ca^{2+}), magnesium (Mg^{2+}), chloride (Cl^-), phosphate (PO_4^{3-}), and bicarbonate (HCO_3^-)] found in cells or the blood.

Electromyogram A graphical tracing of the electrical activity in a muscle at rest and during contraction; used to diagnose nerve and muscle disorders.

Encephalitis An inflammation of the brain, usually caused by a viral infection.

Endemic Referring to a disease that occurs in a specific geographic area.

Endocarditis An inflammation of the membranous lining of the heart's cavities.

Endoscope A medical instrument consisting of a long tube inserted into the body, and used for diagnostic examinations and surgical procedures.

Enteritis An inflammation most commonly of the small intestine.

Epiglottis A flap of cartilage situated at the base of the tongue that covers the opening to the air passages when swallowing, thus preventing food or liquid from entering the windpipe (trachea).

Epiglottitis An inflammation of the epiglottis.

Erysipelas A severe skin rash accompanied by fever and vomiting, and caused by a streptococcal bacterium.

Esophagus The passage through which food moves between the throat and the stomach.

Etiological diagnosis The identification of an illness or disorder in a patient through the identification of the causes and origins of disease.

Febrile Relating to, involving, or typical of, fever.

Filariform larva A threadlike worm of the parasite *Strongyloides* that lives in the soil and burrows into an individual's skin.

Foramen magnum The opening at the base of the skull through which the spinal cord passes into the brain.

Gangrene Local death and decay of soft tissues of the body as a result of lack of blood to the area.

Gastroenteritis An inflammation of the stomach and the intestines, usually as a result of a bacterial or viral infection.

HIV (human immunodeficiency virus) Either of two strains of a retrovirus, HIV-1 or HIV-2, that destroys specific groups of immune cells, the loss of which causes AIDS.

Idiopathic Referring to a disease or disorder that has no known cause.

Inflammation The swelling, redness, heat, and pain produced in an area of the body as a reaction to injury or infection.

Inflammatory response *See* **Inflammation**.

Insulin-dependent diabetes A disorder in which there is no control of blood sugar because of inadequate insulin production; also called type 1 diabetes.

Intubated Referring to a tube inserted through the vocal cords and into the windpipe in order to provide oxygen to a patient's lungs.

Laryngoscope A medical instrument consisting of a short metal or plastic tube fitted with a tiny light bulb, used when examining the larynx.

Lupus (erythematosus) Referring to either of two inflammatory diseases affecting connective tissue. Discoid lupus erythematosus is largely confined to the skin; systemic lupus erythematosus (SLE) affects the joints and internal organs.

Lymphangitis An inflammation of the lymphatic vessels caused by infection that produces conspicuous red streaks under the skin as well as fever.

Malaria An infectious disease caused by a parasite called *Plasmodium* that is transmitted by the bite of infected female mosquitoes and is characterized by recurring chills and fever; also called blackwater fever.

Masseter A muscle in the cheek that moves the jaws during chewing.

Mast cell A type of white blood cell in connective tissue consisting of granules that release histamine and other reactive substances during an allergic reaction.

Meningitis A serious, sometimes fatal illness in which a viral or bacterial infection inflames the meninges, causing symptoms such as severe headaches, vomiting, stiff neck, and high fever.

Meningococcemia The spreading of the meningococcal bacterium *Neisseria meningitidis* through the blood.

Monocyte A type of white blood cell, formed in the bone marrow and in the spleen, that has a well-defined cell nucleus, and consumes large foreign particles and cell debris.

Mononucleosis An infectious disease caused by a virus, producing fever, swelling of the lymph nodes, sore throat, and increased lymphocytes in the blood.

Necrosis The death of cells in a tissue or organ caused by disease or injury.

Neonatal Referring to a newborn child less than one month old.

Obsessive-compulsive disorder A psychiatric disorder characterized by obsessive thoughts and compulsive behavior; for example, continual washing of the hands prompted by a feeling of uncleanliness.

Pancytopenia A severe form of anemia in which the capacity of bone marrow cells to generate red blood cells, white blood cells, and platelets is diminished.

Pap smear A test to detect cancerous or precancerous cells of the cervix, allowing for early diagnosis of cancer.

Penicillin An antibiotic used to treat a wide range of bacterial infections.

Perinatal Referring to the period prior to and after childbirth, usually from around week 28 of pregnancy to one month after birth.

Peritonitis An inflammation of the peritoneum, the membrane that lines the abdomen.

Pia mater The innermost and most delicate of the three membranes that surround the brain and the spinal cord.

Pneumonia An inflammation of one or both lungs, usually caused by infection from a bacterium, virus, or fungus.

Prognosis A medical opinion as to the likely course and outcome of a disease.

Pseudomembranous colitis An infection of the colon often, but not always, caused by the bacterium *Clostridium difficile.*

PTT (partial thromboplastin time) The time it takes for blood platelets to convert prothrombin to thrombin during the process of blood clotting.

Puerperal fever Blood poisoning following childbirth, caused by a streptococcal infection of the placenta.

Pus   The yellowish or greenish fluid that forms at sites of infection, consisting of dead white blood cells, dead tissue, bacteria, and blood serum.

Retroviral drug   *See* **Antiretroviral drug.**

Rheumatic fever   A serious streptococcal disease that causes a sore throat, fever, pain, swelling in the joints, and often damage to the heart valves.

Rifampin   An antibiotic used to treat various bacterial infections, including tuberculosis.

Risus sardonicus   A distorted grinning expression caused by involuntary prolonged contraction of the facial muscles, especially as a result of tetanus.

Rocky Mountain spotted fever   A potentially dangerous infectious disease transmitted by the bite of a tick infected with the microorganism *Rickettsia rickettsii*.

Sepsis   A condition caused by the presence of bacteria, such as *Escherichia coli*, and their toxins in the bloodstream.

Septic shock   A serious medical condition that occurs when an overwhelming infection leads to low blood pressure and low blood flow; leading to decreased oxygen delivery, it can cause multiple organ failure and death.

Sequela   A disease or disorder that follows as a consequence of a preceding disease or injury in the same individual.

Shock   *See* **septic shock.**

Sickle-cell anemia   A group of genetic disorders in which there are too few red blood cells or the red blood cells are deficient in hemoglobin, resulting in sickle shaped red blood cells and poor health.

Sign   An indication of the presence of a disease or disorder, especially one observed by a doctor but not apparent to the patient; for example low grade fever, high blood pressure.

Specific treatment   Refers to a clinician's specific treatment of the diagnosed disease and hopefully affecting the final outcome.

Spinal muscular atrophy   A rare, inherited childhood disease that causes the degeneration of spinal nerves that control body movements.

Spinal tap A surgical procedure that involves drawing spinal fluid using a hollow needle or tube.

Sputum A substance such as saliva, phlegm, or mucus coughed up from the respiratory tract and usually ejected by mouth.

Streptokinase An enzyme produced by streptococci that dissolves blood clots.

Stridor A harsh, high-pitched, wheezing sound made when breathing in or out, caused by obstruction of the air passages.

Strychnine A poisonous nitrogen-containing alkaline compound used as a poison for rodents and medicinally as a stimulant for the central nervous system.

Sural nerve The bundle of fibers that transmits messages near the calf of the leg.

Symptom An indication of some disease or disorder, especially one experienced by the patient; for example, pain, headache, or itching.

Symptomatic treatment Treating the symptoms of an ill patient.

Syndrome A group of signs and symptoms that, taken together, characterize a specific disease or disorder.

Systemic inflammatory response syndrome (SIRS) *See* **sepsis**.

Systemic lupus erythematosus *See* **lupus**.

Tetanus A severe and dangerous infectious disease caused by the bacterium *Clostridium tetani,* usually contracted through a penetrating wound, that causes severe muscular spasms and contractions, especially around the neck and jaw.

Toxin A poison produced by or from bacteria that can cause disease.

Toxoplasmosis A disease caused by the protozoal parasite *Toxoplasma* that is transmitted to humans via undercooked meat or through contact with infectious animals, especially cats.

Trachea The windpipe, which carries air to the lungs.

Trismus A sustained spasm of the jaw muscles, characteristic of the early stages of tetanus.

Trophozoite The active or feeding stage of a protozoal parasite.

Ultrasound An imaging procedure that uses high-frequency sound waves reflecting off internal body parts to create images, especially of a fetus in the womb, for medical examination.

Ventricular tachyarrhythmia A medical condition in which the heartbeat is fast and irregular.

Xenodiagnosis The identification of a parasitic infection that is carried out by allowing a non-infected disease-carrying organism (e.g., mosquito, tick) to feed on an infected person's blood and then examining the original non-infected organism for infection.

Index